LIFE

as

WE MADE IT

ALSO BY BETH SHAPIRO

How to Clone a Mammoth

LIFE
as
WE MADE IT

How 50,000 Years of Human Innovation
Refined—and Redefined—Nature

BETH SHAPIRO

BASIC BOOKS
New York

Basic Books
Hachette Book Group
1290 Avenue of the Americas, New York, NY 10104
www.basicbooks.com

Printed in the United States of America
First Edition: October 2021

Published by Basic Books, an imprint of Perseus Books, LLC, a subsidiary of Hachette Book Group, Inc. The Basic Books name and logo is a trademark of the Hachette Book Group.

The Hachette Speakers Bureau provides a wide range of authors for speaking events. To find out more, go to www.hachettespeakersbureau.com or call (866) 376-6591.

The publisher is not responsible for websites (or their content) that are not owned by the publisher.

Print book interior design by Linda Mark.

Library of Congress Cataloging-in-Publication Data
Names: Shapiro, Beth Alison, author.
Title: Life as we made it : how 50,000 years of human innovation refined—and redefined—nature / Beth Shapiro.
Description: First edition. | New York : Basic Books, 2021. | Includes bibliographical references and index.
Identifiers: LCCN 2021009604 | ISBN 9781541644182 (hardcover) | ISBN 9781541644151 (ebook)
Subjects: LCSH: Biotechnology—History. | Biotechnology—Moral and ethical aspects. | Nature—Effect of human beings on.
Classification: LCC TP248.18 .S46 2021 | DDC 660.609—dc23
LC record available at https://lccn.loc.gov/2021009604

ISBNs: 9781541644182 (hardcover), 9781541644151 (ebook)

LSC-C

Printing 1, 2021

For James and Henry, who found the bits of this book that I read to them to be "medium interesting."

Contents

— PROLOGUE —

Providence

DEEP IN THE HEART OF THE AMERICAN WEST, AN OLD BISON plucks a mouthful of young grass. As her teeth grind the blades, a wolf's howl interrupts the gentle din of the nearby Snake River. The bison raises her head and pauses, suddenly focused. Her ears twitch as she sniffs the air. A quiet moment passes while a mosquito buzzes aimlessly around her head. Sensing no imminent danger, she resumes chewing and returns her gaze to the ground. As she propels herself toward a fresh patch of grass, her movement stirs the several dozen other bison nearby and, together, the herd begins to move slowly southward, silently grazing an unhurried path toward the mountains.

The scene is tranquil, even comforting. A herd of wild bison flourishes in one of the last wild places on Earth, unfettered by all of the stuff that exists outside that wild place. The scene is also hopeful. While it's true that we humans have made a hot mess of our planet, patches of habitat survive where bison can roam, separated from all the world's unnatural tendencies. More importantly, the scene is inspirational. These bison exist because we saved them. By the late 1800s, the millions of bison that once roamed the plains were nearly gone. But bison did not become extinct. Instead, people created safe spaces for their herds to chew cud and raise calves, and designed and passed laws to protect those safe spaces from hunters, poachers, and other threats. Thanks to us, more than 500,000 bison live today in herds across North America.

Above all else, though, the scene is natural. The image of American bison sheltered within the boundaries of North America's first national park is a portrait of unspoiled wildness. This is the way the natural world should be, because this is the way the natural world has always been.

Except for when it wasn't.

THE LAST DECADE HAS SEEN DEVELOPMENT OF POWERFUL BIO-technologies that are at the same time astonishing, encouraging, and pretty scary. Cloning, genome editing, synthetic biology, gene drives—these are words and phrases that promise a different kind of future, but is it a welcome future? On the one hand, technological advance is a good thing. Biotechnology stops us from getting sick, cures diseases that we already have, and makes our food taste better and stay fresh longer. On the other hand, biotechnology creates things that feel weirdly unnatural, like corn with embedded bacterial genes and chickens that lay eggs out of which ducklings hatch.* In fact, it is increasingly difficult to find anything that hasn't been sullied by people in some way. And while scientists race to protect

---

* This is true! It is possible to inject primordial germ cells—the cells that will eventually become either sperm or egg germ cells—extracted from developing duck eggs into chicken eggs. When the chick hatches and reaches sexual maturity, she will have two types of egg cells: some that developed from her own primordial germ cells and others that developed from the injected duck primordial germ cells. If she is artificially inseminated with duck sperm, these will fertilize the duck eggs (they are too evolutionarily different to fertilize the chicken eggs, so no duck/chicken hybrids will form). When those eggs hatch, a baby duck will emerge.

the natural things and spaces that remain, crises like offshore oil spills, rising extinction rates, and emerging infectious diseases demand solutions beyond what our existing technologies can achieve. Should we dig in, embrace the power of modern science, and look ahead to a future where bacteria clean up our messes and where hairy elephants roam Siberian fields while sterilized mosquitoes buzz overhead? Or should we resist this future and stop messing with things before it is too late?

For many, a future filled with human-modified plants and animals is bleak. Engineered microbes, mammothified elephants, and mosquitoes that can't transmit disease would probably benefit people in some way, but creating them just isn't right, and a world that includes them is somehow false. To those who feel this way, there is a tendency to blame science. Thanks to scientists and their twenty-first-century technologies, our world is on the precipice of a metamorphosis beyond which lies a new nature, one created entirely by and for people, and one that is anything but natural. This nervous narrative assumes, however, that humans have only just begun to meddle with nature—that the border between natural and unnatural is obvious and unblurred. History, however, and archaeology and paleontology and even genomics, tells a different story. In studying the past, we learn that people have been shaping the evolution of the living things around us throughout our history. Within the last 50,000 years, our ancestors hunted, polluted, and outcompeted hundreds of species to extinction. They turned wolves into Boston terriers, teosinte into popcorn, and wild cabbage into kale, broccoli, cauliflower, brussels sprouts, and collard greens (to name a few). As our ancestors learned to hunt, to domesticate, and to travel, their actions and movements created opportunities for species to adapt and evolve. Some species survived their encounters with humans, but many did not, and all were transformed in some way. Living things today are as we made them, shaped in part by the randomness of evolution and in part by less random human intent.

Let's consider again the American bison. When people first set foot more than 20,000 years ago on what we now call the North American continent, they may have done so while following these delicious-looking beasts. As time passed, these people developed sophisticated technologies for hunting bison, some of which could kill thousands of animals in a single assault. Only those bison that avoided predation survived. The climate cooled, their habitat deteriorated, and bison populations crashed.

When the ice age came to an end around 12,000 years ago, bison-friendly habitats reexpanded and the herds rebounded. The warmer climate was also good for people, whose populations grew. Vegetation became denser, and people started to use fire to reshape the landscape. They learned to funnel bison into habitats in which the animals could be more easily harvested. Bison adapted to these changes and thrived. People adapted too. Human life came to revolve around the seasonal dynamics of the bison herds. People used bison meat, hides, poop, and bones for food, clothing, fuel, and tools. Trading networks emerged, connecting human populations across the continental expanse.

When Europeans arrived in North America around half a millennium ago, they too developed a taste for bison. As European immigrants pushed westward, the sizable bison herds began to fracture and collapse, divided this time by a growing railway infrastructure and even more people. Wars over ownership of bison and bison ranges intensified, and many people, and many more bison, died. Treaties were signed and then broken and Native Americans suffered disproportionately. Cattle ranching expanded and began to compete with bison for food, space, and water. This time, the changes happened too quickly for bison to adapt. By the turn of the twentieth century, a few bison had been captured and moved to captivity and another few remained in the wild, but the once-massive herds were gone.

Then, around a century ago, people realized that bison were in trouble. Conservation-minded officials passed laws to prevent bison slaughter. Wildlife managers constructed barriers to keep bison safe and, within the confines of these barriers, chose which bison contributed to the next generation. The optimal strategy for bison was no longer to escape, but instead to somehow impress their human captors. Bison adapted and thrived.

Today's biotechnologies allow us to meddle with species like bison more quickly and with higher precision than our ancestors could. Artificial insemination, cloning, and gene editing improve our control over what DNA is passed to the next generation, further heightening the power of human intent as an evolutionary force. To date, these biotechnologies have had the most impact in agriculture. One hundred years ago, a farmer who noticed a piglet that grew larger than its siblings might breed only this more efficient pig, shifting, gradually and over many generations, the efficiency of his drove. Fifty years ago, this same farmer might have collected semen from his fastest-growing boar and used that semen to impregnate his sows, increasing the number of offspring that inherit the fast-growing trait. Today, that farmer could sequence the pigs' DNA to learn what genetic variants distinguish faster-growing piglets from their slower-growing brothers and sisters. The farmer could collect cells from the best of the bunch and clone them, creating embryos that all have the fast-growing DNA variants and transferring these into surrogate sows. The farmer could edit these embryos' DNA directly to create DNA combinations that make his pigs grow even more quickly. The end product of all this meddling is the same: larger pigs and a more favorable bottom line. But today's technologies pay off in years rather than in decades or centuries.

Some new biotechnologies give us powers that our ancestors did not have, and this is where it starts to get tricky. The Enviropig is a pig, and its DNA is mostly pig DNA. But its genome also includes a gene from a microbe and a gene from a mouse. No amount of careful breeding could create the Enviropig, but our technologies can. The Enviropig solves a specific problem in agriculture. The watersheds around pig farms are often badly polluted with phosphorus, a crucial nutrient that farmers add to pig feed and that is mostly excreted in their waste. The two extra genes in the Enviropig's DNA express a protein in pig saliva that breaks phosphorus down into usable form. Enviropigs can be fed less phosphorus than non-engineered pigs (which would save farmers money) and use the phosphorus in their feed more efficiently (which would save the watershed). In 2010, when the Enviropig was submitted for regulatory approval, no one was sure how to proceed. The approval process stalled, and the project eventually ran out of money. Enviropigs solve a problem plaguing one of the largest agricultural industries in the world, but our discomfort with the technologies that created them thwarts the breakthrough that they could bring.

The Enviropig exposes both our uneasiness with transitioning to the next phase of our relationships with other species and the high cost of this uneasiness. Our reluctance has slowed research into both the safety and wider potential of these technologies. We've missed opportunities to adopt bioengineered solutions that might have removed pollutants from habitats, saved populations from extinction, and increased agricultural yield. Our reluctance is, however, understandable. Many of the earliest uses of bioengineering technologies were marred by a lack of transparency, with little or no effort to describe, for example, how bioengineered crops were created or how they differed (and did not differ) from traditional crops. This lack of transparency created opportunities for a handful of vocal extremists to spread misinformation, taking advantage of our natural aversion to risk. Poorly aligned regulatory frameworks and battles over intellectual property have continued to obstruct open discourse about the science behind bioengineered crops. This has left many would-be consumers with an entirely reasonable sour taste when it comes to bioengineered foods.

While bioengineered foods have been on at least the theoretical menu since the mid-1990s, these technologies have more recently been targeting another crisis: the global loss of biodiversity that has occurred as people have taken over the planet. The rate of species extinction today is many times higher than the background rate of extinctions in the fossil record. This is our fault, due to the gradual erosion of the quantity and quality of habitat available to species other than us and our domesticates. While most people agree that we should do something about the extinction crisis, not everyone agrees about what that something should be. Some people want to preserve nature intact by isolating as much as half of the planet from any human inputs. Others believe that direct intervention is the only way to slow the rate of human-caused extinctions. For decades biologists have been removing invasive species by hand, moving individuals between populations and habitats, and introducing proxy species to fill crucial but vacant (usually because of extinction) ecological niches. But today's biotechnologies make it possible to do more. We could, for example, engineer species' genomes to help them adapt to drier soils, more acidic oceans, and more polluted streams. We could create gene drive systems that wipe out invasive species. We could even resurrect extinct species to restore missing ecological interactions and improve ecosystem health. These bioengineered interventions have tremendous potential for conservation but come with additional risks.

In 2017, Helen Taylor and her colleagues surveyed conservation practitioners in Aotearoa/New Zealand to find out how they feel about using genetic engineering as part of their on-the-ground conservation work. The country's government had announced a year earlier an ambitious goal to rid the islands by 2050 of introduced rats, Australian possums, and stoats, all of which are devastating native fauna. The project's timeline is ambitious, and many have looked to bioengineering as a possible route to success. Taylor and colleagues' survey showed, however, that willingness to adopt such strategies depended on which species would be manipulated. For most survey respondents, it was fine to modify the DNA of invasive species, but not OK to manipulate native species in this way. In fact, many respondents admitted that they would rather see native species become extinct than use bioengineering to save them. Their discomfort? They would be playing God. They would be changing deliberately the course of evolution, and they just didn't feel good about taking on this role. Except in the case of invasive species.

Such attitudes are the providence of caretakers. It's time to embrace this role.

In the chapters that follow, I've divided the story of our species' changing relationships with other species into two parts that separate roughly the time before and after the advent of genetic engineering technologies, which many people see as a turning point in our capacity to manipulate nature. Part I, "The Way It Is," progresses through three chronological stages of human innovation: predation, domestication, and conservation. Chapter 1, "Bone Mining," describes my journey from student to professor and the growth of ancient DNA as a research field, introducing how I and others have used DNA preserved in fossils to reconstruct evolutionary history. Chapter 2, "Origin Story," explores what ancient DNA has revealed about our species' origins, including how encounters with our archaic cousins shaped our own evolutionary path. Chapter 3, "Blitzkrieg," describes the gradual spread of humans across the globe and our species' role as a dominant predator, exploring the temporal coincidence of human arrival in an as yet uninvaded habitat and the extinctions of local fauna. Chapter 4, "Lactase Persistence," chronicles our ancestors' transition from hunters to farmers, beginning with the discovery that extinction is not inevitable. To improve the reliability of their next meal, our ancestors developed herding and breeding strategies and began to clear forests to make way for farms. Chapter 5, "Lake Cow Bacon," describes our species' next transition from farmers to managers, which began as booming populations of people and domesticated animals consumed wild habitats and drove to extinction many species within them. This was the birth of the conservation movement.

Today, we rely on the technologies that our ancestors developed during these first three stages of human innovation. Our meddling, however, continues to reshape every living thing around us. Industrialized agriculture is helping to meet the needs of the nearly 9 billion of us alive today, and international laws protect Earth's oceans, air, land, and freshwater ecosystems. But our planet is again reaching a breaking point. There are more of us today than we can feed using existing technologies and, thanks to how we've altered the planet, habitats are changing too quickly for species to adapt and extinction rates are soaring. Once again, however, we have new tools at our disposal, and these new tools allow us to manipulate species at unprecedented rates and in unprecedented ways.

Part II, "The Way It Could Be," explores the biotechnologies of this next stage of human innovation. Chapter 6, "Polled," focuses on how biotechnologies like cloning and genetic engineering impact plant and animal agriculture by allowing us, for example, to engineer domesticated species without the messiness of traditional breeding. Chapter 7, "Intended Consequences," considers how new biotechnologies can protect endangered species and habitats. From cloned mammoths to transgenic ferrets to self-limiting mosquitoes, biotechnologies could speed up the process of adaptation and, in doing so, slow the rate of biodiversity loss and restore stability to declining habitats. Finally, Chapter 8, "Turkish Delight," speculates about what else we might do with our new biotechnologies. No longer limited by traditional species boundaries, will we stick with what we know and continue to make our food, pets, and crops better versions of themselves, or will we reach beyond what is currently imaginable to invent something even better?

Today's biotechnologies differ from those of the past, and this requires us to treat them differently. Our power to change species is greater than it ever has been, and we must recognize, accept, and learn to check these powers. This will not be easy, but it is possible. After all, we are also different today than we were in the past. We have a much greater understanding of how the world works. We have deep knowledge of biology, heredity, and ecology, and we are capable of evaluating risks, communicating across cultures and languages, and sharing intellectual and economic burdens. Critically, we also have tens of thousands of years of experience manipulating nature with the same motivation as today: to create organisms that more efficiently do what we want them to do.

It is a mistake to think of today's biotechnologies as a sudden shift into the realm of controlling nature. We have, after all, been in this role for some time.

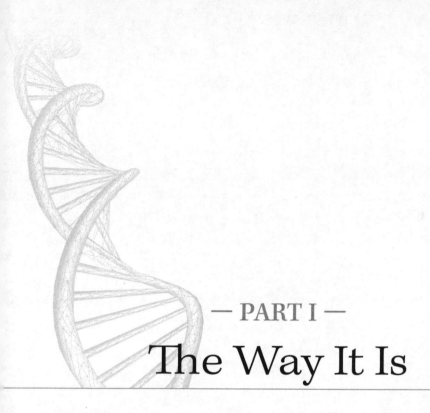

— PART I —

The Way It Is

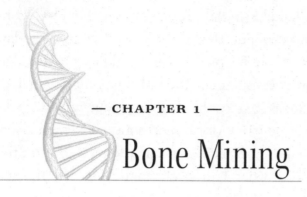

— CHAPTER 1 —

Bone Mining

IN CANADA'S YUKON, IT TAKES LESS THAN AN HOUR IN A four-wheel drive truck to get from the coffee shop on Dawson City's Front Street to the last ice age. Located about 500 kilometers (330 mi) northwest of Whitehorse, Dawson City is a rugged northern town of dirt roads and wooden sidewalks, swinging-door saloons, and rickety buildings heaved onto jaunty angles by the melting earth beneath them. Today, tourism is the chief economic activity in Dawson City. But this has not always been the case. Since the discovery of gold there in 1896, an estimated 15 million troy ounces (which is more than 460,000 kg) have been mined from the nearby Klondike region's extensive creek and river system.

Gold, however, is not the only precious commodity unearthed by Klondike miners. Each year, thousands of ice age fossils wash out of the Klondike's frozen soil during the search for gold, among them remains of mammoths, mastodons, bison, horses, willows, ground squirrels, wolves, camels, spruce trees, lions, lemmings, and bears. The sticks, seeds, bones, teeth, and sometimes entire mummified bodies represent plants and animals that lived in the Klondike at some point during the most recent million or so years.

Since the beginning of the gold rush, scientists have collected and scrutinized the Klondike's fossils, hoping to use them to reconstruct the climate and communities of the recent ice ages. Today, these fossils are a mainstay of my own research, and I try to spend at least a few weeks there every summer. Today, I can point to which of the Klondike's dusty roads are most likely to wash out, which creeks cut through the most productive (for bones) soils, and which layers of volcanic ash can tell us how old a particular fossil is. But on the warm summer day of 2001 when I drove out to the mines for the first time, I knew nothing.

Three of us went out from Dawson City into the Klondike that day: me, and two friends and colleagues, Duane Froese and Grant Zazula. We were all graduate students at the time and each of our research relied on data from the region, but I was the only one of us who'd never been out to the mines. We were in Dawson City for a conference, which meant learning about science during the day and exploring the town's nightlife when the talks were over. We'd been exploring a dingy bar the night before that the locals call "The Pit" when we ran into a miner friend of Duane's who, after several Yukon Golds, invited us out to his place the next day to check out his bone cache. The next morning, as we abandoned the conference and left Dawson City for the mines, I was prepared for the sun. I was prepared for the Klondike's ubiquitous mosquitoes, and for the possibility of running into a bear. I was not, as it turns out, prepared for the muck.

About 20 minutes outside town, we turned off the main highway and began winding along the dusty mining roads in Duane's rental truck. I was struck by the contrast between the Klondike's natural and human worlds. One minute we'd be passing through pristine spruce forest or making our way gingerly over a hopefully not too deep creek, and the next we'd be in the middle of a denuded landscape where bulldozers were scraping away chunks of frozen earth. The road was curvy and washboard-riddled, and my stomach churned as our pickup fishtailed around the turns. When we finally turned onto a long driveway of sorts and slowed down, I was desperate for fresh air. Having been given the middle seat, I reached over Grant to roll down the window, and that is when I learned my first lesson about the Klondike: It *stank*. I gasped as the foul air hit me and, defeated, sank back into my seat. Grant and Duane didn't seem to notice the smell.

A short moment later, we pulled up beside the mine's main shop and parked. Grant and Duane jumped out, but I stayed behind. The stink seemed to be getting worse. I contemplated remaining in the truck while they checked out the bones. But I'd come out because I wanted to see the bones—and the mine—for myself, so I shook the thought. I inhaled one last breath of truck air, opened the door, and emerged into the stench.

As my feet hit the gravel, I steadied myself and looked around. To my right were the shop and a few outbuildings, to my left an outhouse (maybe the source of the foul smell?), a couple of trucks, and what I assumed would amount to several large dumpster-fills of rusty metal contraptions. In the distance, I saw several people, presumably miners, fiddling with what looked like a firehose mounted onto a stabilizing platform. Duane had already started toward them, so I followed, keen to distance myself from whatever was creating that smell.

Oddly, the closer we got to the miners the more intense the stench became. I looked at Grant and pinched my nose in disgust. Ahead of us, a powerful generator kicked on, assaulting yet another of my senses. I groaned audibly and kicked a large rock, which hit Duane on the back of his boot.

Duane looked back at me. "What?" he yelled over the generator, seemingly oblivious to the smell.

Grant laughed. "It's her first time!" he reminded Duane.

"Ah," nodded Duane, who'd turned back around and was now squinting into the sun, trying to see whether his friend was part of the group by the firehose. "The smell, eh? What do you think this muck is made of?" he asked of no one in particular.

"That's dead mammoths," Grant instructed, smirking. "And dead trees and grass and other shit that's been rotting since the last ice age."

That made sense. Frozen organic debris tens of thousands of years old, exposed suddenly to the summer sun, could certainly generate an unpleasant odor.

"And glacial silt," added Duane. "You'll want to be careful."

The three of us continued in the direction of the now-running firehose contraption, me acclimating to the smell and the noise, and Duane waving his arms over his head and shouting. The miners spotted us and reduced the flow of the water, causing the generator to change pitch. Interpreting this as an invitation, Duane hurried toward them for a chat. Grant and I waited where we were, scanning the freshly exposed muck for signs of ice age life.

Almost immediately, I spotted the tip of a bison horn sticking out of the frozen soil near the base of the exposure. I jabbed Grant in the ribs and pointed to it excitedly. He smiled, impressed (I assumed) with my bone-mining skills, and motioned for me to go grab it. Stoked to have found my very first ice age fossil, I headed toward the bluff. I tiptoed over a shallow stream of water running from the blasting site and leaped over a puddle that had formed in a small depression. That was when I learned my second lesson about the Klondike: One must tread delicately. As yet unaware of this rule, my indelicate landing left me ankle-deep in the muck and sinking. I panicked. I yanked one foot upward. While that foot didn't move, the other, thanks to the extra pressure, sank deeper into the muck. I yanked my foot upward again. This time the foot came up, but the boot stayed behind. I teetered, my socked foot hovering above the wet muck, then lost my balance and fell backward. With both feet, both hands, and my bum planted in the stinky quicksand, I looked back to Grant for help, only to find him doubled over laughing at the predicament he'd gotten me into.

"I told you to be careful!" Duane shouted from beside the fire-hose, where he and the miners were all watching me, smiling, and shaking their heads.

After I finally unstuck myself from the muck (an endeavor that involved removing both boots, losing one sock, and becoming one with the stench of decaying dead things thousands of years old; also, I now know, a rite of passage for working in the region), we made our way back to the mining shop to look through their cache of bones. They were mostly bison bones, which pleased me as I was currently studying ice age bison, but there were also horse bones, mammoth bones and bits of tusks, caribou bones and antlers, and the odd bear and cat bones. We were told to take the bones back to the museum in Whitehorse, so we labeled each bone and recorded in our field books what species they were, the date of collection, and the name of the mine. Using a battery-powered drill, I took small samples from several of the bison bones, from which I would extract DNA after returning to my lab in Oxford. Then we thanked the miners, packed up our notes, and loaded the bones into Duane's truck, readying them for transfer to Whitehorse.

HOW IT ALL STARTED (FOR ME)

When I began my graduate research in 1999, I hadn't intended to study bison. I wasn't thinking about bison as I slipped nervously through the halls of Oxford University's zoology department for the first time. No thoughts of bison surfaced when I found the desk at which I would sit for the next five years. I had no particular interest in bison as a child, nor would I meet a bison until several months later, when I used a Dremel saw to cut into a 30,000-year-old bison bone (yes, that counts). I'm ashamed to admit, in fact, that my first bison-related thought was not particularly bison-friendly, but instead an awkward mental race to imagine an appropriately kind but negative response to my soon-to-be supervisor's suggestion, "Why don't you work on bison?" Fortunately for my career, he followed up with "If you take this project, you can go to Siberia." How could I say no?

These were early days of the scientific field known as ancient DNA. The field had been born some 15 years earlier, when scientists working in Allan Wilson's research lab at the University of California, Berkeley, recovered and sequenced DNA from a small piece of muscle tissue snipped from 100-year-old preserved remains of a quagga, which is an extinct type of zebra. Their discovery that DNA was sometimes preserved in dead organisms ignited a scientific frenzy. Teams aiming to sequence DNA from other extinct species formed in labs around the globe, racing to recover preserved DNA from mammoths, cave bears, moas, and Neanderthals. As competition raged to publish the oldest DNA and DNA from the most unusual specimens, little attention was paid to authenticating the most spectacular results. By the mid-1990s, dinosaur DNA and DNA attributed to insects entombed in fossilized amber had been published in reputable scientific journals. The excitement was palpable, but there was a problem. While some published ancient DNA sequences could be validated, none of the extremely old DNA sequences were real. In fact, most (but not all) DNA sequences purportedly older than a few hundred thousand years have since been identified as contaminants, sometimes from microbes, sometimes from people, and sometimes from what the researchers had for lunch. These were the dark days of ancient DNA.

In 1999 as I began my career in ancient DNA, the field was beginning to find its footing as a serious scientific discipline. Scientists had learned that ancient DNA tends to be broken down into tiny, chemically damaged fragments, and that ancient DNA experiments are often contaminated with undamaged DNA from living organisms, such as from the person doing the work. A few institutes and universities had invested huge sums of cash during the late 1990s to develop meticulously clean labs for ancient DNA research. The scientists leading these labs proposed strict protocols for performing ancient DNA research, including working only in these sterile environments, soaking everything in bleach to destroy potentially contaminating DNA, wearing sterile suits, booties, gloves, hairnets, and face masks to avoid contaminating ancient samples, and distrusting results produced by competitor labs. These measures had the secondary effect of limiting the number of labs that could compete in the quest for the most interesting, most ancient DNA.

When I toddled into Oxford and ancient DNA, blissfully unaware of the emulous world into which I was stepping, the lab there was only beginning to take shape. Alan Cooper, its director and my soon-to-be boss, had just arrived from Allan Wilson's group in Berkeley, where he and many influential early figures in ancient DNA were trained. Alan had secured a clean space for ancient DNA research at the Oxford University Museum of Natural History and hired Ian Barnes to work in the lab as a postdoctoral scholar. When I agreed to join them, we were three.

One might imagine that in a relatively new research field in which only a few labs were participating, I might have my choice of research topics. Not so in ancient DNA. By 1999, all the taxa had been divided up among labs, and the most exciting ones—carnivores, ancient humans, and anything else likely to pique the interest of scientific editors and journalists—had been claimed. Svante Pääbo (also from Allan Wilson's group) and Hendrik Poinar, at the newly established Max Planck Institute for Evolutionary Anthropology in Leipzig, Germany, claimed mammoths, giant ground sloths, humans, and Neanderthals. Bob Wayne at University of California, Los Angeles, claimed dogs, wolves, and horses. Ross MacPhee at the American Museum of Natural History claimed muskoxen. Alan claimed bears and cats, which had been subsequently subclaimed by Ian, and bison, which nobody seemed too bothered about.

Although I wasn't exactly enamored with bison, I found ancient DNA to be unambiguously appealing. During an undergraduate summer field program in geology, I'd become fascinated by how Earth processes shaped living systems. I was particularly intrigued by the visible scars on the landscape from successive advances and retreats of massive glaciers during the Pleistocene epoch—the geological period of time that spanned most of the last several million years. I imagined how each advancing glacier must have reset the living systems in its path, driving extinctions, creating new combinations of species, and providing opportunities for evolution. The most recent ice age also coincided with the first major influx of people into North America, which no doubt added to the slow-motion biological upheaval precipitated by glacial retreat, not dissimilar to the slow-motion biological upheaval taking place today. In fact, I had chosen Oxford precisely to study this connection between the past and the present—to learn paleontology and evolutionary biology, two of the university's strengths, and to combine these with my training in geology and ecology. Until meeting Alan, I had not heard of ancient DNA, but its potential to reveal how recent ice ages affected the evolution of life on Earth was immediately obvious. If I could learn to extract and analyze ancient DNA, then I could trace evolutionary changes as they were recorded in DNA during past periods of biological upheaval. I could learn lessons from the past that would be relevant to protecting species and ecosystems today. Yes, I was a bit giddy with enthusiasm, but ancient DNA was *so cool*.

There were problems with this plan. I had a sum total of zero experience in molecular biology. I'd never handled a pipettor or extracted DNA. I had no idea what particular bits of DNA I should study. I didn't know where or how to get fossils from which I might extract DNA. And I knew nothing about bison.

Not to be deterred, I decided to begin in the library. I swore an oath to neither drink tea in the stacks nor set fire to the library's books (as required to get my Oxford University library card) and set out to learn about bison. I found a surprising number of relevant books, few of which appeared to have ever been opened. For several weeks, I sat in the dark damp cold of the library's basement, resisting the surprisingly potent lure to get warm by drinking hot tea or setting books on fire, and learning about bison, which turned out to be way more interesting than I'd imagined them to be.

WHAT IS A BISON?

Known as *tatanka* or *pte* in the Lakota language, *tł'okjjeré* in Dene, and many other names in the languages of the people who lived with them for millennia, the animals now best known as bison were given their first English name by sixteenth-century Europeans, who called them buffalo. Today, the word *buffalo* connotes unflappability and intimidation, which seems appropriate for these hulking and headstrong animals. In the sixteenth century, though, it referred to a puffy leather coat, or buffe, that the colonists envisaged making from the animal's hide. As if being named for outerwear weren't bad enough, *buffalo* wasn't even a unique moniker, but one that Europeans ascribed to all kinds of newly discovered (by them) animals that they imagined turning into coats. The so-many-buffalos problem caused consternation among European taxonomists, who as a profession demand meticulous curation when it comes to names. By the middle of the eighteenth century, wrangling over taxonomic priority was infused with a touch of common sense and the number of buffalo was reduced to three: North American buffalo, African buffalo, and Asian buffalo. Then, in 1758, Carl Linnaeus officially designated the American buffalo *Bison bison*, and taxonomists breathed a collective sigh of relief. The American buffe, they declared, should officially be called a bison. And sometimes we remember to call them that.

Bison are relative newcomers to North America. While mammoths and horses have been part of the North American fauna for millions of years, bison, which evolved in Asia some 2 million years ago, appeared in the local fossil record more recently. They entered North America by crossing the Bering Land Bridge, which is named for the Bering Sea under which the bridge is currently submerged, which is itself named for Vitus Jonassen Bering, a Danish explorer and mapmaker who made the same voyage some hundreds of thousands of years later by boat.

While nineteenth- and early twentieth-century paleontologists did not know precisely when bison crossed the Bering Land Bridge, they did have some clues. They knew, for example, that the land bridge was available only during the coldest parts of the Pleistocene ice ages when, because much of the planet's fresh water was frozen into ice, the sea level was lower than it is today. When exposed, the Bering Land Bridge formed a continuous corridor of unglaciated habitat through which animals could move freely. Although not on the same schedule or in the same direction, mammoths, lions, horses, bison, bears, and even people all used the Bering Land Bridge to disperse between continents. Because the land bridge was only exposed intermittently, paleontologists knew that bison must have entered North America during a cold stage. They knew that bison must have made the voyage relatively recently. Most bison bones found on the continent are not yet mineralized, which means they are probably not very old. A few from warmer climates where mineralization happens more quickly, however, are partially mineralized, which confused things. What paleontologists needed was a way to measure the age of the bison fossils directly. This became possible in the 1950s with the advent of a new technology called radiocarbon dating.

Radiocarbon dating can reveal how long ago an organism died. The technology takes advantage of the fact that organisms take up carbon from the atmosphere as they grow and use it as a building block for making bones, leaves, and other body parts. In fact, they take up two different isotopes of carbon: carbon-12, which is stable, and carbon-14, which is a radioactive isotope created when cosmic rays hit Earth's upper atmosphere. Carbon-14 is unstable and decays back to carbon-12 with a half-life of 5,730 years. After an organism dies, that organism stops taking in new carbon from the atmosphere but the carbon-14 in the biological remains continues to decay to carbon-12. This means that the ratio of carbon-14 to carbon-12 in the remains will decrease at a known rate over time. By measuring this ratio, we can learn how many years it has been since the organism stopped taking in new carbon-14. This tells us when the organism died and therefore how old the fossil is.

Radiocarbon dating revolutionized paleontology, but the method has limitations. Most importantly, radiocarbon dating can only be used to learn the ages of relatively young ancient remains. After 50,000 years or so, too little carbon-14 is left to measure accurately, and the approach can only reveal that the sample is older than this limit.

When radiocarbon dating was used to estimate the ages of the oldest bison bones in North America, most were younger than 50,000 years but some were too old to date. This meant that bison entered North America sometime prior to 50,000 years ago. And that was where things stood for more than half a century, until ancient DNA finally solved the mystery, with the help of a volcano.

In 2013, Berto Reyes, a geologist from the University of Alberta, was working in the far north of Canada's Yukon when he found a frozen bison foot bone sticking out of an equally frozen cliff. The cliff was part of Ch'ijee's Bluff, a geological exposure near the remote settlement of Old Crow. The bone was wedged just above a thick volcanic ash called the Old Crow Tephra, in a prominent layer of dark brown soil filled with sticks, plant roots, and other organic debris—the type of deposit that forms during the warm periods between ice ages. Above the thick dark layer in which the bone was wedged was a layer of fine gray silt—the kind of deposit that accumulates during ice ages. Because geological layers accumulate over time, Berto knew that the bone was from an animal that lived during a warm period before an ice age. This was a clue as to how old the bison foot bone was. But there have been as many as twenty warm periods and ice ages during the Pleistocene, so how could he know which of these warm periods he had discovered?

This is where the volcano comes in.

When a volcano erupts, fragments of rocks, crystals, and glass are ejected into the upper atmosphere as ash. The ash is swept into wind currents and carried away from the volcano, sometimes as far as thousands of kilometers. Eventually the ash falls to the ground, settling like snow across the landscape into layers that range from microscopic to several meters thick. These ash layers are gradually covered by sediment that builds up over time, swept around and deposited by wind and rain and other normal processes. Many thousands of years later, a river might cut through the accumulated earth, exposing a canyon wall that looks as if a white blanket has been accidentally sandwiched between layers of otherwise normal-looking soil. That white blanket is the ash and a marker in time: everything below the ash—all the soil and any bones, trees, or other organic material—was deposited before the volcano erupted, and everything above the ash was deposited after the eruption.

Layers of volcanic ash, also called tephra, are common in ice age deposits in Alaska and Yukon. Two nearby volcanic fields, the Aleutian arc/Alaska Peninsula region and the Wrangell Volcanic Field in southeastern Alaska, are sources of tephra-producing eruptions. Because each volcano has a slightly different elemental signature, the tephra falls produced when they erupt provide geological fingerprints that can link ash discovered across the region to the same eruption. Importantly, the glass particles within the ash can also reveal when each eruption occurred, thanks to a method similar to radiocarbon dating that measures the radioactive decay of uranium-238, which, because it has a longer half-life, can provide dates for eruptions that happened as long as 2 million years ago.

The Old Crow Tephra (the tephra above which Berto's bison foot bone was found) was estimated by geologists to have been deposited around 135,000 years ago. We knew from the composition of the soil layers above the tephra that the bison lived during a warm period before a more recent ice age. And we knew from the geological record that there was only one window of time between 135,000 years ago and the most recent ice age during which the climate of the northern Yukon was sufficiently warm to support woody plants: around 119,000 to 125,000 years ago. This must have been when Berto's bison lived.

This foot bone that Berto found is the oldest plausibly dated bison fossil in North America. We and others have looked for bison in lots of older deposits, including several fossil sites that sit below (and are therefore older than) deposits of Old Crow Tephra. Horse bones, mammoth bones, and bones from other ice age animals are common in these older sites, but none of these sites include bison. When we extracted ancient DNA from Berto's bison bone, we found that it fell within the genetic diversity of all extinct and living North American bison, indicating that all these bison are part of that same Bering Land Bridge–crossing lineage. Berto's bison was among the earliest bison to live in North America.

Bison probably dispersed across the Bering Land Bridge around 160,000 years ago, when the bridge was exposed during the cold period that predated Berto's bison. As the climate warmed and grasslands expanded, bison dispersed eastward and southward and spread throughout the continent. This spread is documented as tens of thousands of bison fossils of various shapes and sizes found from Alaska to as far south as present-day northern Mexico and from west to east across nearly the entire continent. The most extreme of these was the aptly named giant long-horned bison, *Bison latifrons*, which had horns that spanned more than 210 centimeters (nearly 7 ft) from tip to tip. Long-horned bison were more than double the size of their northern contemporaries and, judging from their co-occurrence with other species that thrived during the warm interglacial period, were at least 125,000 years old. In fact, long-horned bison were so distinct from other bison that some paleontologists believed they were a different species that separately crossed the land bridge. Oddly, however, no long-horned bison fossils have been found north of the midcontinent despite the exceptional fossil record in that part of the world. If they were descended from a separate invasion across the Bering Land Bridge, long-horned bison must have run across the northern continent too quickly to be recorded in the fossil record.

As a graduate student, I knew that I could answer this question if I could extract DNA from a giant long-horned bison. Unfortunately, long-horned bison lived both very long ago and during a warm period, both of which are bad for DNA preservation. For years, I tried and failed to recover ancient DNA from remains of long-horned bison. I graduated, moved on to other projects, and gave up on the idea. Then, on October 14, 2010, Jesse Steele accidentally ran over a mammoth with his bulldozer, and I was presented an opportunity.

Steele was part of a crew expanding a water reservoir serving the community of Snowmass Village, Colorado, which is known for its proximity to some of the best skiing in the Rockies. When he plucked the odd giant rib from his bulldozer's teeth, Steele had no idea that he had just discovered one of the richest deposits of ice age fossils in North America. Work on the reservoir halted temporarily as a team led by the Denver Museum of Nature and Science and the United States Geological Survey descended on the site. For eight weeks during the summer of 2011, hundreds of museum staff and volunteers and dozens of scientists who, like me, couldn't stay away donned bright yellow safety vests and gleaming white hard hats and began to excavate. In the end, we collected more than 35,000 plant and animal fossils. We recovered dozens of long-horned bison, as well as mastodons, mammoths, ground sloths, camels, horses, and smaller animals like salamanders, snakes, lizards, river otters, and beavers. Preservation was exceptional. Leaves from 100,000-year-old sedges and willows were still green as they were removed from the mud. Pieces of ancient driftwood were recovered that were up to 20 meters (65 ft) long. Shells from beetles, mollusks, and snails retained many of their original bright colors. I had every reason to be hopeful for preserved long-horned bison DNA.

In the end, though, we were able to recover DNA from only one long-horned bison bone. This best-preserved bison was in a layer of the ancient lake that was deposited around 110,000 years ago. Its DNA was badly degraded, but we painstakingly pieced together a sequence to add to our larger data set. When we ran the new analyses, the result was unambiguous: long-horned bison were not genetically different from other bison, despite their distinctive morphology. Giant long-horned bison were not a different species but instead an ecomorph—a lineage that looks unique because of adaptations to a different environment. Their near doubling in size compared to the Chi'jee's Bluff bison was presumably thanks to ample resources in the North American midcontinent during the warm interval in which they lived.

As the planet cooled and the grasslands disappeared, so did the long-horned bison. By 90,000 years ago, all bison were small and North America was again plunging into an ice age. At one of our field sites in Yukon, we've found thousands of bison bones associated with a different volcanic ash layer, the Sheep Creek Tephra, which was deposited around 77,000 years ago. The abundance of bison at this site, in terms of both the absolute number of bones and their commonness relative to other animals like mammoths and horses, suggests that bison populations in Yukon were massive at this time. In fact, the period from 77,000 years ago until the onset around 35,000 years ago of the coldest part of the last ice age should probably be renamed "peak bison."

Throughout the peak bison period, bison continued to disperse between the cold habitats in the North and the warmer American midcontinent. Migrating herds would have encountered and bred with other herds, generating new morphological and ecological diversity. This diversity was a professional playground for nineteenth- and early twentieth-century paleontologists, who relied on the tiniest of details—slight differences in horn curvature, distance between horns, or shape of the eye socket—to declare a fossil a Never Before Seen And Therefore Obviously New Species. Remains, especially skulls, were measured, sketched, and measured again. These measurements were the tea leaves through which new species were identified, scientific papers written, and paleontological fame achieved.

I love stories about this period of the bison bone rush. My friend Mike Wilson, an expert on bison taxonomy, shared many amusing stories about dodgy taxonomical goings-on around this time. His stories all convey a sort of paleontological rule-oriented laissez-faire. For example, to decide whether a bison fossil represented a new species (yay!) or an already named species (boo!), a paleontologist might take the skull and place it on the ground with the nose pointing forward and the horns to the left and right, and then measure the ratio of the length to width of the skull at the base of the horns. This measurement would then be compared to measurements from already described bison species to determine whether this bison could be something never before seen. One might imagine that as more and more fossils were measured the number of new species discoveries would slow down. Not so. In fact, at some point during the bison bone rush, people started documenting horn position with the nose of the skull pointing to the left rather than to the front, such

that what was once length was now recorded as width,* presumably
to maintain the high rate of new species "discoveries."

As a consequence of the bone rush, bison from this period have
been assigned dozens of scientific names: *Bison crassicornis, Bison oc-
cidentalis, Bison priscus, Bison antiquus, Bison regius, Bison rotundus,
Bison taylori, Bison pacificus, Bison kansensis, Bison sylvestris, Bison
californicus, Bison oliverhayi, Bison icouldgoonforeveri, Bison youget-
thepointus.* However, by the end of the twentieth century, some paleontologists were convinced that there was in fact only one bison
species in North America, and so I set about to test this hypothesis
using ancient DNA. With the permission and assistance of many
patient museum curators, I collected tiny pieces of bones from fossils
that had been assigned these and other names. I visited museums
across North America, spending days at a time in behind-the-scenes
rooms packed with movable shelving units onto which thousands of
fossils were stacked, finding, identifying, and drilling small chunks
out of hundreds of bones. On breaks, I would emerge, disoriented,
into brightly lit exhibit halls, clutching my temporary security pass.
Museum visitors who happened to be nearby were no doubt amused
by the sudden sighting of a captive scientist, deep-red creases on my
face delineating my temporarily shed dust mask and white streaks of
bone dust in my hair. I also never seemed to emerge into the same
part of the museum, as if my hosts were using me to enhance their
visitor experience. I was happy to oblige.

I brought these bits of bison bone back to Oxford, where I extracted and sequenced their DNA, and then compared the recovered

---

* This is an approximation of what Mike told me. It may instead have been
that the nose was first pointed to the left and then to the right, or some other
orientation-based mishap. Regardless, this is a terrible morphological character
with which to identify species, and not only because the rule was not followed
correctly. Bison horns are shaped by a combination of factors that have nothing
to do with what species they belong to, including how healthy the animals were
as their horns were developing and how much time they spent fighting with
other bison during their lifetime.

DNA sequences to one another and to sequences from living bison. The ancient bison were genetically different from present-day bison. Specifically, their genomes harbored much more diversity than persists today. This told me that ancient bison populations were enormous. However, I found no evidence that ancient bison genetic diversity was partitioned among the species to which the fossils had been assigned. There was, according to the DNA, only one bison species in North America. So what should this species be called? The answer to that question is surprisingly easy. The rules of taxonomy (of which there are many) dictate that if a single species has been assigned multiple names, then the first name that they were assigned has priority. North American bison are therefore *Bison bison*. All of them.

A TURN FOR THE WORSE

Things began to go badly for North American bison around 35,000 years ago. Until then, the ice age had been pretty mild as far as ice ages go. In Beringia, which is the name given to the geographic region that extends from the Lena River in western Siberia to the Mackenzie River in Canada's Yukon and includes the now-submerged Bering Land Bridge, bison habitat had been plentiful. Annual rainfall was too low for the region to become glaciated but sufficiently high to support a rich steppe grassland that was ideal for bison. But as the climate cooled and rainfall decreased, grasses began to be replaced by less nutritious shrubs. Horses, which could get by eating shrubs, boomed temporarily as they outcompeted grass-dependent bison. Their success was also short-lived, however, as the climate continued to deteriorate and even the shrubs began to disappear.

By 23,000 years ago, around the peak cold period of the last ice age, both bison and horses in Beringia were in serious trouble. Habitat was scarce, and two massive glaciers, the Cordilleran ice sheet along the eastern foothills of the Rocky Mountains and the Laurentide ice sheet across the Canadian shield, had merged in present-day

western Canada, cutting off movement between Beringia and potentially better habitat to the south. This ice barrier remained in
place for nearly 10,000 years.

The coldest part of the last ice age was not a bison-friendly time
anywhere in North America. Not only did the grassland habitat
nearly disappear, but a new predator appeared with sights firmly fixed
on bison. This predator, which had more recently crossed the land
bridge from Asia, walked upright on two legs and could fling sharp
points over long distances. These were the first people to set foot
on the continent, and bison had never before encountered predators that hunted as they did. The usual predators of bison—wolves
or bears or big cats—might take one or two animals, probably the
youngest, oldest, or sickest, in a single successful hunt. People, however, worked together, and could take dozens of animals or more at
a time. Rather than target the weakest in the herd, people took the
largest, the healthiest, and the fattest bison. As human populations
grew and the hunt for bison intensified, bison herd structures were
disrupted. Breeding seasons came and went and fewer cows meant
fewer calves, and bison populations began to decline. At the same
time, bison habitat was becoming increasingly sparse, and surviving
herds were forced into isolated and shrinking patches of grassland.
Unfortunately for the bison, people also knew where those patches
of habitat were.

While paleontologists can infer the ice age decline in bison by
counting the number of bison bones that date to this period, much
stronger evidence of their slow demise comes from ancient DNA.
By the time the ice age came to an end and the world started to
warm up, the once large and connected Beringian bison population
had been reduced to several small and geographically isolated herds,
each eking out a separate living in the patches of grassland that
remained. These remnant populations hung on, a few of them for
thousands of years, but the time of plenty was over. Individuals living in these patches were genetically similar to one another—a sign

that the populations were very small—and rarely moved between patches. By 2,000 years ago, the last of the northern populations of bison was extinct.

Bison that had been trapped to the south of the ice when the Laurentide and Cordilleran glaciers merged were also in trouble. Humans were there, too, at least toward the end of the ice age, living in scattered populations and developing new tools, some of which were designed specifically to kill bison. By 13,000 years ago, bison south of the ice were reduced to just one or perhaps a few surviving herds. Today, all living bison descend from this southern population. If it hadn't been for the survival of a few bison to the south of the ice sheets, bison, like mammoths, giant bears, North American lions, and so many other charismatic ice age animals, would be extinct.

TEMPORARY RECOVERY

As the climate warmed out of the last ice age and today's warm period—a new geological epoch called the Holocene—began, the grasslands returned to the North American midcontinent in full force. Mammoths and horses were already or about to be extinct, which meant that bison had less competition for this expanding ecological niche, and by 10,000 years ago bison were back and thriving. Millions of (closely related) bison had expanded across the plains (these would later be called plains bison) and into forests farther north (these would later be called wood bison). The Early Holocene was an ideal time to be a bison in North America.

People were doing well, too, of course. By the time the grasslands returned fully to the North American plains, humans were settled nearly everywhere across the continent. These early people of North America were exceptionally creative when it came to killing bison. Armed with spears and bows and arrows, people drove bison into snowdrifts, backed them into canyons and corrals, and ambushed them as they tried to cross rivers and lakes. They chased bison into frozen water, attacked them at drinking holes, and forced them to run off the edges of steep cliffs, with death or serious injury resulting either from the fall or from being crushed as animals fell on top of one another. At these mass kill sites, called "buffalo jumps," it was not unusual for several hundred animals to be run off a cliff in a single drive.

Communal bison hunting was an important part of the social lives of early North American people. Communal kills meant communal harvests, bringing normally dispersed groups together to make the most of the (sometimes literal) mountains of dead bison. During these communal harvests, families were reunited, achievements were celebrated, marriages were arranged, and political decisions were made. Bison leather was shaped into footwear, canoes, and teepee covers, and horns were made into rattles and cups. These physical items, as well as songs, stories, dances, and art, preserve memories of the more than 14,000 years of interactions between people and bison in North America. People relied on bison, and bison helped to shape human evolutionary history.

Bison, too, were adapting to life with humans. By the Early Holocene, bison were smaller in size than their ice age ancestors. In his book *Frozen Fauna of the Mammoth Steppe*, Dale Guthrie, a paleontologist from the University of Alaska, Fairbanks, attributed the reduction in bison size to interactions with people. The main ice age predators of bison were lions, giant bears, and saber-toothed cats, which hunted alone or in small groups. To avoid being eaten, a bison might confront the attacker, or even mount a counterattack using his large sweeping horns. After these animals went extinct, gray wolves and humans became the main predators of bison. When a wolf pack or communal hunting party attacks, the best survival strategy is to run away. These new hunting pressures would have favored the evolution of smaller nimbler bison. Canadian biologist Valerius Geist agrees, pointing out that humans probably also imposed strong selection pressures against large and brave bulls, as these would have been the most likely bison to confront a spear-wielding human and therefore the most likely to be killed. By around 5,000 years ago, which was probably 15,000 or so years after they first encountered humans, North American bison looked pretty much as they do today, with a body mass of around 70 percent that of their ice age ancestors.

Humans may also have driven the reduction in bison size indirectly. Sustained by an endless supply of bison meat, humans established their permanent settlements in the best habitats; after all, an armed human will always win a territorial battle. Bison were banished to the sidelines where the vegetation was not as abundant or nutritious, and where smaller bison that required fewer resources had an advantage over their larger brothers and sisters. Bison adapted and hung on, although in smaller bodies and numbers than before.

Then there was a respite, at least for bison. Around 500 years ago, Europeans arrived in North America, and smallpox, whooping cough, typhoid, scarlet fever, and other diseases swept across the continent, ravaging indigenous populations. With so much death, many of the long-established human settlements disappeared, reducing hunting pressure on bison. By the middle of the eighteenth century, historical records count as many as 60 million bison in the North American plains. This final success would not last.

In addition to diseases, Europeans brought horses back to North America. Horses evolved in North America millions of years ago but went extinct on the continent at the end of the last ice age, surviving only in Europe and Asia. The reintroduction of horses by Spanish explorers in the sixteenth century was bad news for bison. By the eighteenth century, everybody—colonists and indigenous people—had discovered horses to be particularly handy for rounding up and killing bison. Europeans also brought guns, which were second to horses in their contribution to the increased speed and accuracy with which bison could be slaughtered.

The pace of European colonization quickened in the early 1800s, and along with it the bison hunt. Waves of westward-expanding colonists identified bison as good to eat and easy to sell. Trappers and traders sent hundreds of thousands of bison hides back to the East Coast each year. Train companies offered the nineteenth-century version of in-flight entertainment, which was to allow riders to shoot at bison as the cars passed through the plains. Most unfortunately, bison came to be seen by government and military leaders as resources of the enemy, where the enemy was any Native American who refused to give up hunting bison, be relocated to reservations, and take up farming. Military leaders and politicians viewed the complete destruction of bison as the only solution to this intractable "problem," and encouraged the killing of any and all bison, regardless of how many treaties were broken in the process. The dozens of large herds of the mid-eighteenth century declined to just two by 1868, one in the northern plains and another in the southern plains, divided by the railroad. An economic downturn in 1873 compelled more trappers, or buffalo runners, to the prairie, hoping to turn bison hides into cash. Their success flooded the market, reducing the value of each dead bison, making it necessary to kill even more bison to make a living. Herds disappeared from the plains, leaving in their place piles of bones and rotting unused carcasses (minus the hide). By 1876, bison had disappeared entirely from the southern plains. By 1884, fewer than 1,000 bison were alive in North America.

SALVAGE

Rumblings of discontent about the human role in the impending bison extinction started to be heard in the early 1800s. It wasn't until 1874, however, that the first legislation to protect bison—an act to prevent their "useless slaughter" that was supported by both houses of the US Congress—was passed. Unfortunately, President

Ulysses S. Grant refused to sign the act into law. In 1877, the Canadian government passed the Buffalo Protection Act, which hoped to achieve the same goal, but then also refused to enforce it. Finally, in 1894, President Grover Cleveland signed into law the Lacey Act, which was an "Act to protect the birds and animals in Yellowstone National Park and to punish crimes in said park." This act protected the only remaining free-ranging plains bison in North America (a free-ranging population of wood bison survived in western Canada). A census of the Yellowstone population in 1902, eight years after their protection was codified, counted fewer than 25 bison.

Fortunately, bison had protectors other than the law. In the 1870s and 1880s, private citizens began to take interest in bison, particularly in their commercial potential. They rounded up whatever wild bison they could find and established six private herds from a total of around 100 bison. Because we're counting, if we add those 100 bison to the 25 or so wild bison that survived in Yellowstone, this means that all living plains bison descend from around 125 individuals. This was the second near extinction of bison in fewer than 15,000 years.

In 1905, the American Bison Society was founded as a conservation organization with the stated aim to save the American bison from extinction. Theodore Roosevelt, himself an enthusiastic bison hunter, became the society's first honorary president. Two years later, the society led the second animal reintroduction to wildlands in the United States (the reintroduction by Spanish explorers of long-extinct horses was the first), when they shipped 15 bison from the herd established at the Bronx Zoo to a ranch in Oklahoma. A year later, the society successfully petitioned Congress to establish the National Bison Range in Montana, which they populated in 1909 with bison purchased from private landowners using funds raised by the society. Within the next five years, similar strategies were used to establish bison herds in Wind Cave National Park in South Dakota and Fort Niobrara Wildlife Refuge in Nebraska. The herds thrived. Bison were saved.

TODAY

Two types of bison live in North America today: plains bison and wood bison. Both descend from bison that survived the near extinctions 13,000 and then 150 years ago. Plains bison, which are officially *Bison bison* of the subspecies *bison*, making them *Bison bison bison*, inhabit the grassland plains, whereas wood bison, which are taxonomically distinguished as *Bison bison athabascae*, inhabit more northerly and mountainous regions of the continent. Wood bison are slightly larger than plains bison and said to have less hair—finer beards and manes and less hairy forelegs. The thin evolutionary line separating wood and plains bison was blurred in 1925 when the Canadian government moved 6,000 plains bison from Buffalo National Park in central Alberta to Wood Buffalo National Park in northern Alberta in an effort to relieve grazing pressure in the central province. Lacking evolutionary barriers to stop them, wood and plains bison interbred, and today there are probably no genetically "pure" wood bison herds. Nonetheless, governments protect wood and plains bison separately in an effort to preserve their unique characteristics and diversity.

Five hundred thousand (or so) bison live today in herds that range in size from ten to thousands of animals. Bison are large animals—they can weigh up to a ton—with a placid demeanor and poor eyesight, but they are glorious creatures. They are generally calm, although they can run as fast as a galloping horse and, if spooked, jump nearly two meters (six feet) vertically from a standstill. In regional and national parks with captive bison herds, rangers warn visitors of the perils of interacting with bison, and yet every year a few people ignore these warnings and manage to get hurt or worse. These are, after all, wild animals, and not simply hairy versions of domesticated cattle.

Bison are a conservation success story. After nearly becoming extinct during the late nineteenth century, today's herds are healthy and stable, a strong and profitable market exists for their meat, fur, and skins, and in 2016 they were declared by President Barack Obama as the national mammal of the United States. Bison populations within North America are pleasantly sustainable.

What does success as pleasant sustainability look like? Today, most bison are privately owned and raised as livestock. These bison are selectively bred for traits that make them manageable on the ranch and profitable at the market, such as tameness, high fertility, rapid growth, and efficient feeding. Cattle genes are found in many (possibly all) herds, thanks to deliberate crossbreeding of the two species during the early twentieth century by ranchers who wanted livestock with the temperament of cattle but the hardiness of bison. Breeding within and between herds is carefully managed, and several large nonprofit and for-profit organizations offer services including DNA typing, testing for diseases, and facilitating trades. Within the protected confines of their pastures, these bison don't have to compete with other grazers. They don't have to worry about predation by wolves or bears. They can't migrate to track quality forage as seasons change. The traits and genes that help these bison survive are no longer the same as those that made their ancestors successful. Are these bison still wild?

Some bison—around 4 percent of living animals—live in conservation herds. These herds altogether occupy less than 1 percent of the range that bison once used. Bison in conservation herds are not selectively bred for commercial purposes, but they are no less heavily managed than bison in private herds. Like commercial herds, conservation herds graze within fenced areas, which protects them from diseases, predators, and other scary situations. They are culled annually to limit their population growth, animals with less than ideal temperaments are removed from the herd, and the herd's sex and age structure are optimized to control reproduction and reduce the possibility of escape.

Most conservation herds also have some cattle ancestry, which leads to questions about their suitability for conservation. For example, around 50 percent of the bison in the Santa Catalina Island conservation herd off the coast of California have cattle mitochondrial DNA—a type of DNA that is inherited down the maternal line. When James Derr, a bison biologist at Texas A&M University, studied the Santa Catalina herd to learn whether having cattle DNA changed these bison in some way, he found that bison with cattle mitochondrial DNA were smaller and shorter than those in the herd with bison mitochondrial DNA. These results suggest that there may be consequences to the Santa Catalina herd from having so much cattle ancestry. Are these still wild bison?

Given results like Derr's, herd managers struggle with how to protect conservation herds. Should cattle ancestry be selected against and removed, or should mixed ancestry be embraced as an opportunity to increase genetic variation in a species that, thanks to their near extinction, are all genetically similar? At the same time, herds are slowly accumulating herd-specific genetic variation, both through local adaptation, which is good, and inbreeding, which can be bad. This presents a difficult choice for herd managers. On one hand, moving individuals between herds can prevent inbreeding from making the entire herd less fit. On the other hand, that movement may overwrite local adaptations that allow bison to survive in new and changing environmental conditions. Regardless of what choice they make, the herd managers determine the evolutionary fate of the American bison.

EVOLUTION WITH INTENT

The story of the North American bison encapsulates our species' takeover of the evolutionary process. For nearly 2 million years, bison evolved in the absence of people. They adapted to the coming and going of ice ages. When it was cold, bison with thicker fur were more likely to be strong and healthy and able to escape if a predator attacked. When it was warm, perhaps the opposite was true. Predators weeded out the oldest and the least fit bison, and those bison that survived reproduced. Bison spread across the Northern Hemisphere, thriving and declining along with their grassland habitat.

Then humans arrived, bringing tools that could be redesigned and adjusted more quickly than bison could evolve to escape. For more than 14,000 years, people hunted bison for food and for sport. Twice during this time, bison nearly became extinct. Their first recovery was down to luck: the Early Holocene was an ideal climate for expanding grasslands and bison had few herbivorous competitors for these vast resources. It was, however, only temporary.

When bison were nearly gone for the second time, people decided to save them. Wildlife managers took over where evolution once reigned. They decided which bison survived and reproduced, even breeding some bison with cattle. Government officials created bison reserves and passed laws to protect those reserves. Today, our governments consider each herd to be a separate worthy entity and protect them as such, despite the fact that all living bison descend from fewer than 125 ancestors. Some bison herds are designated for conservation while others are not. Managers and breeders move herds between habitats and individuals between herds. Scientists analyze bison DNA to help managers decide which individuals should be bred, what genes are the correct genes, and how much cattle DNA is too much cattle DNA. Managers cull bison to limit overgrazing, vaccinate them to prevent disease, and fence them in to exclude predators. And within the confines of those fences, we stare at them from the comfort of our cars, mesmerized and reassured by their enduring wildness.

The evolution story of the North American bison is one in which humans are deeply entwined, as we are with nearly all (or perhaps all) the organisms alive today. Our role in shaping bison changed from that of a predator to that of a protector, with intermediate phases in which we learned to manipulate and manage these wild animals to suit particular needs. In the chapters that follow, I will explore these roles and the transitions between them more deeply, drawing examples from my own and other research to reveal how we shape and reshape species' evolutionary trajectories as we move along our own evolutionary path.

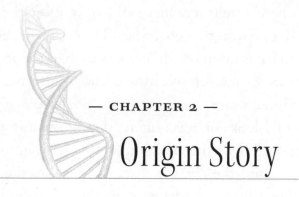

— CHAPTER 2 —

Origin Story

*H*OMO SAPIENS. THAT'S US—OR AT LEAST THE LATIN BINO-
mial that Linnaeus ascribed to us in 1758. That name places
us in the genus *Homo* and the species *sapiens*. Today, we are the only
Homo around, but that hasn't been true for long. Our evolutionary
cousins, *Homo neanderthalensis*, lived alongside us until about 40,000
years ago, when they became extinct. Well, mostly extinct, since
a lot of their DNA is still passed down from one generation to the
next, although now in bodies labeled *sapiens* rather than *neanderthalensis*. Which, if one thinks about it, kind of questions the utility of
"species" as a discrete designation. Clearly, though, we are different
from other species, including from our archaic cousins. So what does
it actually mean to be *Homo sapiens* or, to use our common name, to
be human?

Nearly every species that has ever lived is now extinct. Most
species stick around for somewhere in the range of half a million
years to 10 million years, which is not a particularly long time on
the scale of life's history. But the news isn't entirely bleak, because
evolution.

As soon as a species emerges, it begins to change. Individuals reproduce and pass on copies of their genomes to the next generation. The copying process isn't perfect, though. In fact, DNA copying errors cause every child to have around 40 differences in their genome compared to the genomes of their parents. Most of these differences, which are often called mutations, have no effect. Some, though, change the way the child looks or acts, differentiating them from others in their generation. This is the variation on which evolution acts. Some mutations change the child in a way that makes them less likely to find food or a mate. Others improve their chances of surviving and reproducing. The mutations in the genomes of the most successful children—which evolution defines simply as those who grow up to have the most kids—will become more and more common in the species over time.

Species evolve in this way until one of two things happens. Eventually, either the last individual dies and the species becomes extinct, or a taxonomist decides that the species has changed enough to deserve a new name. When the latter happens, the old species is gone, but its DNA is still passed on, only now through a host that is called something else. Is this also an extinction? Before rushing to answer, keep in mind that the billions, maybe even a trillion, species alive today all descend from a single microbe that lived some four billion years ago. That microbe is both extinct and survives in every living thing.

New species emerge by chance. A plant floats to a distant island and takes root, founding a new population. A river changes course, slicing a once large population into two smaller ones. A few individuals discover a new habitat, move there, and begin to reproduce. These are all opportunities for genetic isolation. If the new population remains isolated, it will evolve along a separate trajectory from other populations, accumulating a different suite of mutations. Eventually, the new population may become so distinct that a new species will have been born.

Time is another form of genetic isolation. Because new mutations accumulate in every generation, the genome of any individual
born thousands of generations from now will include a suite of accumulated mutations, some of which may be incompatible with the
genomes of individuals alive today. As a consequence, species have
finite life spans. They are born and they die at the pace of evolution.

The rules of evolution are simple. Mutations accumulate. Most
of the time it is a simple game of chance that decides whether those
mutations are passed to the next generation. Sometimes, though,
an individual is born with a genetic variant that makes them more
likely to survive and reproduce—in other words, more fit—in the
environment in which they are born. These mutations have a better chance of sticking around into the next generation. Over time,
lineages diversify and adapt as mutations arise and either increase in
commonness or disappear. And extinction happens.

For most of our own evolutionary history, our lineage was
no different from any other. Our ancestors were among those in
their populations that survived and reproduced. Over billions of
generations, our ancestors' genomes accumulated mutations and
our lineage adapted. Climates changed, habitats shifted, niches
emerged and disappeared. Our lineage became animals, then
mammals, then primates, then apes. And then our ancestors figured out how to break the rules. They learned to work together
to overrule chance, to help others rather than allow those less
fit among them to die. They learned to fashion their surround-
ings rather than be changed by them. They learned to direct
evolution—to determine both their own evolutionary trajectories and those of the species with which they interacted—rather
than be subject to its whims. And while paleoanthropologists
still don't fully understand where or when or how this happened,
this is how we became different, unquestionably, from every
other species that lives or has ever lived on Earth. This is what it
means to be human.

THE RISE OF US

Around 40 million years ago, during the Eocene, a lineage of monkey-like primates dispersed from the warm and stable climate of Southeast Asia to colonize Africa. The Eocene became the Oligocene, and ongoing movement of tectonic plates lifted mountains in the Great Rift Valley and shifted climate and weather patterns. Glaciers formed at the poles and the planet became cooler. The once dense African jungles began to dry out, transforming some habitats into savannas and deserts. The monkeys adapted to these changes, evolving new strategies to find food and a safe place to sleep.

Around 26 million years ago in the Great Rift Valley, the first fossils ascribed to an ape appeared in the fossil record. *Rukwapithecus fleaglei* is known only from a jawbone, but its teeth are sufficiently distinct from other primates for paleontologists to classify it as something new. *Rukwapithecus* and other apes had larger bodies and larger brains than Old World monkeys. They also lacked a tail, indicating that they had evolved some other mechanism to maintain balance and move between trees.

The climate continued to change, and the apes were an evolutionary success. Around 18 million years ago, a new kind of ape with strong jaws and thick teeth appeared in the fossil record. This ape, *Afropithecus*, could chew up and extract nutrients from foods that were inaccessible to their weaker-jawed, fruit-eating cousins, foods like thick-husked plants and hard-shelled nuts and seeds. This ability to exploit a wider range of foods may have been the evolutionary advantage that allowed *Afropithecus* or their descendants to leave Africa and expand into Asia and Europe, where they thrived, at least initially.

During the Early Miocene, which began around 23 million years ago, the European climate was subtropical and ideal for the mostly fruit-eating apes. However, as the Miocene progressed, the planet cooled. European jungles began to be replaced by woodlands and endless summers gave way to seasons. As habitats shrank and resources became scarcer, European apes adapted and diversified. Some lineages evolved an upright posture, powerful grasping fingers, strong wrists, and larger brains—adaptations to moving quickly and strategically through the shrinking European jungles. Then around 10 million years ago, these larger and presumably more versatile apes returned to Africa. These apes are probably our ancestors.

A truly bipedal lifestyle is a key characteristic of the hominins, which is the name often given to the subgroup of apes that includes our own genus, *Homo*. The first hominin to habitually walk upright was *Australopithecus*. *Australopithecus* evolved around 4 million years ago in eastern Africa, and most paleoanthropologists believe that *Australopithecus* was a direct ancestor of our lineage, *Homo*. Proof that *Australopithecus* walked on two feet comes from sets of fossil footprints left in volcanic ash 3.66 million years ago that were rediscovered in 1976 in present-day northern Tanzania. Australopith fossils uncovered throughout the African continent reveal a highly adaptable, diverse group that lived in an increasingly dry and unpredictable climate. Most australopiths had brains that were about 35 percent of the size of a modern human brain. All australopiths had dexterous wrists and hands. The remains of Lucy, a 3.2-million-year-old *Australopithecus afarensis* discovered in Ethiopia's Afar region, had robust legs, as expected for a bipedal hominin, *and* robust arms. Her species, it seems, was equally agile in the trees and on the ground. When savannas began to expand around 3 million years ago, some australopiths evolved thick molars that allowed them to extract nutrients from the savanna's fibrous and gritty grasses. Unlike later *Homo*, however, australopiths never entirely abandoned the woodlands for the savannas but preferred either dense forests or mosaic habitats of woodland and savanna where trees probably provided refuge from hungry carnivores.

With bipedalism came some advantages. A bipedal ape can raise itself higher to scan across longer distances than can an ape that stands on all fours. A bipedal ape can reach higher-growing fruit and spot more distant prey. A bipedal ape also has two free limbs with hands and wrists already capable of manipulating small objects. No longer required for standing or running, these free limbs can be co-opted for other purposes. As our ancestors evolved to walk upright, our lineage began to innovate, and to teach one another to repli-cate these innovations. They developed language and cooperation that differentiated them from other apes. They began to shape items from their environment into stone tools—weapons that would im-prove their predatory efficiency.

There would be consequences.

THIS LAND IS OUR LAND

The climatic effects of the Pleistocene ice ages, which began around 2.6 million years ago, were most marked at the poles, but equatorial Africa did not escape the influence of the growing and shrinking ice sheets. When glaciers advanced and sea levels dropped, the African continent became both cooler and drier. When glaciers retreated, the arid conditions across Africa waned. The amplitude of these changes increased twice during the Pleistocene, first around 1.7 mil-lion years ago and again around 1 million years ago. Both times, the corresponding arid cycles in Africa intensified, making it less likely that the moist-adapted ecosystems could recover before the next on-set of aridity. Gradually, African jungles were replaced by woodlands and then by dry savannah grasslands.

Both the flora and fauna of Africa adapted to the Pleistocene climate regime. Species that required wet climates went extinct or evolved to tolerate long periods of aridity and an increasingly unpre-dictable climate. The fossil record reflects this turnover. Rates of ex-tinction increased at the onset of the Pleistocene for many African lineages, including bovids (relatives of present-day cattle, goats, and sheep), suids (relatives of present-day pigs and warthogs), and monkeys. There was of course variability both regionally across Africa and among taxonomic groups. But the pattern is consistent: when Pleistocene glaciations began around 2.6 million years ago, local and regional climates changed dramatically, and the extinction rate increased.

Among herbivores, that is.

Among carnivores, the first few hundred thousand years of the Pleistocene were rather uneventful in terms of extinctions. The rate of carnivore extinctions did eventually tick upward, but not before around 2 million years ago—more than half a million years after the onset of Pleistocene glaciations. Also, once their rate of extinctions increased, carnivores fared worse than other groups of African animals, suffering more extinctions per species than herbivores suffered.

This pattern is puzzling.

A simple explanation might be that some lag time was necessary for carnivores to be affected. As the primary consumers of African plants, herbivores probably took a direct hit once the flora started to change. Only when the herbivores started to go extinct would carnivores have been challenged by a lack of resources. While this explanation makes some sense, half a million years is a long lag— too long—between the onset of herbivore extinctions and when the carnivores begin to feel the effects of those extinctions.

Another related possibility is that carnivores, given their position in the food chain, were more sheltered from Pleistocene climate changes. As herbivore species went extinct, other herbivore species may have increased in abundance to fill the newly emptier planteating niche, providing sufficient food, albeit in a different package, for carnivores. In this scenario, carnivores may not have reached an extinction tipping point until the intensity of the climate changes reached some threshold and the total number of herbivores started to decrease. This explanation also makes sense, but the intensity of glaciations reached its first stepwise increase 1.7 million years ago, which is 300,000 years after carnivore extinction rates rise. This is also probably not the whole story.

Another possible explanation comes from the archaeological record. In 2011, the West Turkana Archaeological Project began to investigate an early human occupation site to the west of Lake Turkana. One day during the first year, several members of the team took a wrong turn and found themselves in an unexplored area. To their astonishment, they found what members of the team described as "unmistakable stone tools" lying on the ground's surface. They began an immediate excavation and, by the end of the following year, had recovered more than 100 pieces of stone tool artifacts and more than 30 fossils of a lineage named *Kenyanthropus* (or sometimes *Australopithecus*) *platyops*. *Kenyanthropus* went extinct around 3.2 million years ago. This particular site dates to 3.3 million years ago, making these the oldest stone tools known and, as of now, the only stone tools that predate the Pleistocene ice ages. These tools are proof that our ancestors and their cousins were evolving to be efficient predators and processors of meat just as things were about to get (climatically) difficult for their prey. Could it have been our ancestors that finally tipped the scales against the carnivores?

Stone tools are observed with increasing frequency in the archaeological record after the onset of the Pleistocene. Bones of a 2.6-million-year-old, cattle-like animal from a site in Gona, Ethiopia, show signs of cut and scrape marks, suggesting that early hominins had begun to process meat and extract bone marrow. Three other sites in Ethiopia and Kenya that date to around 2.35 million years ago confirm the importance of stone tool technologies to early *Homo*. At one of these, a site called Lokalalei in West Turkana, processed bones were recovered from bovids, suids, equids, an extinct rhinoceros, and several large rodents, as well as from reptiles and fish. Our ancestors were, by 2.35 million years ago, competing with other African carnivores for prey.

HANDYMEN (AND -WOMEN)

The acerbity of taxonomic battles involving bison pales in comparison to those involving *Homo*. In the case of *Homo*, however,
the problem is not that there are too many fossils, but that there
are so few. Every new fossil find is met with equal parts fanfare and
handwringing, as the early history of our genus is unwritten, then
rewritten, then agitated over and re-rewritten. Historically, *Homo*
fossils have often been secured in private museum collections, safe
from both accidental damage and the scrutinizing gaze of cantanker-
ous competitors. Happily, however, the newer generation of paleoanthropologists are rejecting secrecy in favor of open scholarship,
even releasing instructions to print precise 3-D replicas along with
reports of new fossil finds.*

Homo evolved in Africa at some point after 3 million years ago,
but precisely when, where, and how many *Homo* lineages coexisted
during this period remain contentious. The oldest known fossil
that *might* belong to *Homo* is a 2.8-million-year-old jawbone, the
Ledi-Geraru mandible, which was discovered in the Afar region in
present-day Ethiopia. The Ledi-Geraru fossil is a partial left mandible with six teeth, three of which have broken crowns. Since no

---

* In 2015, paleoanthropologists Lee Burger and John Hawkes described a
newly discovered species of extinct human, *Homo naledi*, which lived in South
Africa until as recently as 260,000 years ago. Within 12 hours of the team
announcing their discovery, instructions for printing 3-D models of many of
the *Homo naledi* fossils could be downloaded for free. This approach was markedly different than what had been the norm in paleoanthropology, where an-
nouncements of new fossil finds would generally be made many years before
raw data were made available for study or casts were available for purchase.
Today, instructions for printing 3-D models of several *Australopithecus* and
Homo fossils that are held in collections at the National Museums of Kenya
and the Turkana Basin Institute field stations are available for download from
AfricanFossils.org, a nonprofit site that makes the data available via the Creative Commons Attribution-Noncommercial-ShareAlike License.

other part of the skeleton was preserved, Ledi-Geraru's taxonomic assignment hinges on whether the teeth are more similar to teeth from later fossils assigned to *Homo* or to teeth from fossils assigned to *Australopithecus*. Perhaps unsurprisingly, the Ledi-Geraru teeth fall somewhere in between these extremes: they are a bit small for *Australopithecus* and their shape falls outside the range of known *Australopithecus* teeth, and yet they also don't really fall within the range of the earliest *Homo*. The Ledi-Geraru mandible may be that of an early *Homo*, but it also may not be.

In 1960, Jonathan Leakey, the eldest son of the famous paleontologists Louis and Mary Leakey, discovered a lower jaw and top of a skull that belonged to what appeared to be a child while excavating in Olduvai Gorge in Tanzania. The senior Leakeys had begun this excavation nearly 30 years earlier following the discovery of primitive stone tools in the gorge. One year earlier, Mary uncovered a skull from a young adult with a protruding face and a small brain—not, they thought, likely to be the maker of the tools. So they kept looking. Johnny's child, as the new fossil came to be known, was clearly different. Over the next three years, other elements belonging to the same species as Johnny's child were discovered: a child's wrist and hand bones, an adult foot, a partial skull with well-preserved small teeth, and another skull with both jaws. A team of archaeologists and paleoanthropologists including Louis Leakey, Phillip Tobias, Michael Day, and John Napier studied the bones and came to the same conclusion: These elements were from a species that was remarkably similar to modern humans and certainly distinct from the australopiths from southern Africa. This species was probably the maker of the stone tools. They named the new species *Homo habilis*, the toolmaker. *Homo habilis*, which lived from at least 2.4 to 1.7 million years ago, stood upright at 100–135 centimeter (3 ft, 3 in–4 ft, 5 in) tall, walked on two feet, and had a brain that was much larger than earlier *Australopithecus*, but still less than half that of modern humans.

In 1972, a team of archaeologists led by Richard Leakey, who was the second son of Louis and Mary Leakey, and Meave Leakey, Richard's wife, discovered a skull on the eastern shore of today's Lake Turkana. The skull had characteristics of the genus *Homo* but was different from *Homo habilis*. The Leakeys named this fossil *Homo rudolfensis*, after the old name for Lake Turkana, Lake Rudolf. Over the next few years, several more fossils were found in Kenya that appeared to be from *Homo rudolfensis*. Since then, disagreement has arisen (unsurprisingly) as to whether the discovered *Homo rudolfensis* fossils all belong to the same species, whether some should be reassigned to *Homo habilis*, or whether any of them should be assigned to *Homo* at all. Regardless of their specific taxonomy, these fossils date to the same interval as *Homo habilis* and several species of australopith, indicating that multiple lineages of upright-walking human-like primates lived in Africa by 2.4 million years ago.

By 1.8 million years ago, people ascribed to the genus *Homo* were living in present-day Dmanisi, in the Republic of Georgia. These people have been called *Homo erectus*, *Homo ergaster*, and *Homo georgicus*, among other names. They made crude stone tools and while their brains were only slightly larger than those of *Homo habilis*, they were taller. This increase in height and perhaps locomotion may have been the adaptation that enabled the first human exodus from Africa, although this is far from settled.

By 1.6 million years ago, *Homo erectus* were established in Kenya, Tanzania, and South Africa, and by 1 million years ago had dispersed north and east to far eastern China and into Indonesia. By 780,000 years ago, descendants of *Homo erectus*, sometimes called *Homo heidelbergensis*, were settled in present-day Spain, and by 700,000 years ago were as far north as present-day England. These later hominins were big-brained, tall, and lanky, ranging in height from 145 to 185 centimeters (4 ft, 9 in–6 ft, 1 in) and weighing between 40 and 68 kilograms (88–150 lbs). They had highly variable morphologies, even within the same fossil site, similar to modern humans.

Also like modern humans, *Homo erectus* packed an ecological punch. As the first obligate bipeds, *Homo erectus* were exquisitely adapted to chasing down prey. They were endurance hunters, able to run longer distances and cool off more quickly than animals that move on all fours. As intelligent, cooperative, and creative people, they could collaborate on the hunt, increasing their chances of success. Like their ancestors, *Homo erectus* made and used stone tools, but *Homo erectus* refined this skill, refashioning and improving their design over time. As they evolved, *Homo erectus* began to make and use other kinds of tools as well, like wooden spears fashioned for thrusting and perhaps for throwing. They also colonized islands reachable only by boat, and built settlements with separate spaces for butchering meat, working with plants, and sleeping. Although *Homo erectus* did not speak like modern humans do, these complex behaviors imply a level of cooperation, teaching, and learning that could only have been feasible with complex language. *Homo erectus* had begun to challenge the rules of evolution.

LOVE THY NEIGHBOR

By 700,000 years ago, lineages belonging to the genus *Homo* were distributed from the southern tip of Africa northward and across Europe and Asia. They were sophisticated tool users with large brains and dexterous hands, and they had begun to manipulate the world around them. After that, and with apologies to paleo-anthropological scholars for the important details that I am certain to omit, the fossil evidence points to a scenario something like this: *Homo erectus* gave rise to *Homo heidelbergensis*, who spread across Africa and into Europe and lived from around 700,000 years ago until around 200,000 years ago. *Homo heidelbergensis* gave rise to our cousins, *Homo neanderthalensis*, or Neanderthals, by 400,000 years ago and probably in Europe or the Middle East, and to us, *Homo sapiens*, in Africa by 300,000 years ago. Today, every described *Homo* lineage apart from our own is extinct. The dates of the most recent remains of many late-surviving lineages coincide with evidence that our own lineage had turned up in their habitats. This uncanny co-incidence of events also has a simple interpretation: our ancestors killed off all the other human lineages in Africa, left Africa for Europe where they killed off the Neanderthals, and then spread across the rest of the world killing off whatever remnant populations of nonsapiens human they encountered. But we'll return to that part of the story later.

Although my summary is missing many of the finer details, that was essentially the big picture of relatively recent human evolution when the field of ancient DNA was first established. Given our species' selfish tendencies, it's not surprising that Neanderthals and early *Homo sapiens* were among the first targets for ancient DNA research. What these early studies found, however, surprised everyone.

The first Neanderthal DNA sequence was published in 1997. This study, like most genetic research that focuses on Neanderthals, was led by Svante Pääbo, who at the time was a professor at the University of Munich but is now director of the department of genetics at the Max Planck Institute for Evolutionary Anthropology in Leipzig. In 1997, Pääbo and colleagues published the sequence of a small fragment of Neanderthal mitochondrial DNA. Mitochondrial DNA was a common target of early ancient DNA studies for a few reasons. First, every cell has thousands of copies of the mitochondrial genome (mitochondria are organelles found outside the cell's nucleus and have their own genomes) but only two copies of the nuclear genome. This means that mitochondrial DNA is simply more likely than nuclear DNA to survive in fossils. Second, mitochondria are passed down the maternal line, which makes their evolutionary history simple to interpret. The mitochondrial DNA that Pääbo published in 1997 was distinct from all mitochondria present in modern humans, which suggested (like the fossil record) that Neanderthals and humans evolved along separate paths. Fragments of mitochondrial DNA recovered from several other Neanderthal bones were soon added to the evolutionary tree. All these data pointed to the same conclusion: Neanderthals and humans were closely related but clearly different evolutionary lineages that had been evolving separately for at least a few hundred thousand years.

For nearly a decade, that was all ancient DNA revealed about the relationship between Neanderthals and humans. Then, in the early 2000s, new approaches to sequencing DNA made it economical and practical to try to sequence a Neanderthal nuclear genome. In 2006, a team led by Ed Green, who was at the time a postdoctoral researcher in Pääbo's group, published a proof-of-concept paper showing that it would soon, probably, be possible to map out the entire Neanderthal nuclear genome. While their data set was small—the sequences covered only around 0.04 percent of the Neanderthal nuclear genome—it established the approach that all of us working in ancient DNA use today.

In 2010, Ed Green, Svante Pääbo, and others collaborated to produce a complete draft sequence of the Neanderthal genome and, for the first but certainly not last time, ancient DNA rewrote human evolutionary history. With this first draft genome, the team confirmed that Neanderthal populations and modern human populations separated around 460,000 years ago, approximately when the fossil record suggests that the earliest hominins with typical Neanderthal morphologies appeared in Europe. However, the data also delivered a surprise. Some fragments of what could now be identified as Neanderthal DNA were present in modern human genomes. This could only be explained if our cleanly branching evolutionary tree were not cleanly branching after all—if the lineages leading to Neanderthals and modern humans had first separated and then, later, come back together again.

Since 2010, technologies to extract, sequence, and assemble ancient DNA into genomes have continued to improve, and genomes have been assembled from around a dozen Neanderthals that lived between 39,000 and 120,000 years ago in Europe and Siberia. These genomes revealed that Neanderthal populations were small and tended to be geographically isolated from other Neanderthal populations. They've also confirmed what the very first Neanderthal genome revealed: our two lineages had a deeply intertwined evolutionary history. The genomic data contain proof that the two lineages came into contact more frequently than the fossils record and, when this happened, they often mated and exchanged genes. As a consequence, the genomes of most humans alive today include some Neanderthal DNA.

As this DNA-based discovery was threatening the fossil record's supremacy as Keeper of Our Evolution Story, two other findings from ancient DNA knocked the fossils squarely off their pedestals. First, a genome sequenced from an approximately 80,000-year-old finger bone found in Denisova Cave in Siberia turned out to be neither a Neanderthal nor a modern human but a distinct and hitherto undiscovered hominin. Second, a genome sequenced from a 420,000-year-old hominin bone found in a cave in Spain turned out to be, well, confusing.

In 2008, Russian archaeologists excavating Denisova Cave in the Altai Mountains found a partial finger bone that belonged, many tens of thousands of years earlier, to a young girl. The bone was no larger than a coffee bean, but it contained remarkably well-preserved DNA. When compared to human and Neanderthal DNA, her DNA revealed that the girl belonged to an entirely new species of hominin that was until then completely absent from the fossil record. Pääbo's team named this new human lineage "Denisova," after the cave where the finger bone was found.

Analyses of the girl's genomic data have revealed that more than 390,000 years ago, soon after the Neanderthal lineage diverged from the lineage leading to us, a group of Neanderthals left Europe and began to spread across Asia. These people are the Denisovans. That is, of course, not the end of the story. Ancient DNA from subsequently discovered remains has shown that later, perhaps hundreds of thousands of years later, Neanderthals again dispersed from Europe into Asia and, for a time, also hung out in Denisova Cave. While they were there, they shared more than just space with the Denisovans. In 2018, Pääbo's group described from remains recovered in Denisova Cave a female that lived around 90,000 years ago whose mother was a Neanderthal and whose father was a Denisovan. The bone they used to describe this woman was 3 centimeters long and 1 centimeter wide and entirely undiagnostic, at least from a morphological perspective.

Denisovans may have been widespread during the Late Pleistocene. Most fossil evidence of their existence is from Denisova Cave: four teeth and lots of bone shards identified as Denisovan using DNA or protein-based methods. In 2019, however, protein sequences extracted from a 160,000-year-old jawbone found in a cave on the Tibetan plateau provided the first tangible evidence that Denisovans lived outside the Altai Mountains. Even before these fossils were characterized, however, genomic data from modern humans had shown that Denisovans (or maybe not the Denisovans per se, but hominins related to Denisovans) were widespread. Today, around 5 percent of the genomes of people native to Oceania can be traced back to admixture with Denisova-like hominins that is thought to have occurred as modern humans dispersed through present-day Papua New Guinea.

The second blow to the fossil record came a few years later. A team led by Matthias Meyer, also from Pääbo's group, recovered ancient DNA from a 420,000-year-old hominin bone discovered in a cave system known as Sima de los Huesos in the Spanish Atapuerca Mountains. The Sima de los Huesos hominins comprise 28 nearly complete skeletons whose taxonomic identity has always been a point of contention. Juan-Luis Arsuaga, a paleoanthropologist at Universidad Complutense de Madrid, spent decades excavating and studying these remains, and was convinced that they were early Neanderthals. Others argued that the skeletons show diagnostic traits of *Homo heidelbergensis*. When Meyer's team pieced together a mitochondrial genome and parts of a nuclear genome from one of these skeletons, their results didn't vindicate anyone. Mitochondrially, the Sima de los Huesos individual was more similar to Denisovans than to later Neanderthals. Its nuclear genome, however, looked like that of a typical Neanderthal.

What was going on?

One possibility, although just a hypothesis, is that the mitochondrial DNA found in later Neanderthals originated in a lineage other than Neanderthals. In 2017, scientists discovered fossils from modern humans at a site called Jebel Irhoud in Morocco that dates to around 315,000 years ago. If modern humans had already colonized habitats this far to the north by this time, they may have ventured even farther northward, perhaps running into and mating with Neanderthals along the way. Neanderthal populations were small at the time, which makes it possible that part of the Neanderthal genome, such as the mitochondrial DNA, could have been replaced with DNA from these early modern humans. This hypothetical scenario could explain why the Sima de los Huesos Neanderthal has a different mitochondrial sequence than later Neanderthals: the Sima de los Huesos Neanderthal has a version of the "original" Neanderthal mitochondrial DNA, whereas later Neanderthals have a version of mitochondrial DNA that evolved in the lineage leading to modern humans and then got into the Neanderthal population after they mated with these early modern humans. If true, this would imply that our lineage left Africa more than once within the last several hundred thousand years.

So, let's recap the story according to ancient DNA.

Estimates from the *genetic* record indicate that sometime around 460,000 years ago and probably in Africa, a population that was ancestral to modern humans and Neanderthals split into two lineages. One of these remained in Africa and evolved into modern humans, and the other probably dispersed northward into Europe, where it evolved into early proto-Neanderthals. Sometime before 420,000 years ago (the age of the Sima de los Huesos bones), the proto-Neanderthal lineage split into two. One stayed behind and evolved into western Neanderthals, and the other dispersed eastward, evolving into Denisovans (one might even call them eastern Neanderthals). After the Neanderthal/Denisovan split but before 125,000 years ago (when the mitochondrial sequence that is evolutionarily more similar to ours is observed in a Neanderthal in Germany), modern humans may have moved out of Africa much earlier than the fossil record suggests, mated with western Neanderthals, and then disappeared. At least some western Neanderthals inherited mitochondrial DNA from these early modern humans. Neanderthals carrying this early modern human mitochondrial DNA then dispersed eastward, where they ran into and mated with Denisovans. Sometime after 70,000 years ago, modern humans moved out of Africa and into Europe again, this time with more enduring success. There, they encountered and mated with Neanderthals again. Modern humans then dispersed eastward, mating with and replacing Neanderthals and Denisovans along the way.

Why was there so much hanky-panky in our evolutionary history? The answer is superficially simple: If they could, then why not? There was no barrier to gene flow when these populations encountered each other, so exchanging genes . . . happened. But there also may have been an evolutionary benefit to this lack of reproductive barrier. Populations that have been established in a location for a little while may have all sorts of adaptations to living there, ranging from immunity against local pathogens to adaptations to local climate and diet. By mating with distantly related evolutionary cousins, our ancestors may have inherited genes that helped them to survive, even thrive, in their new environments.

If all this mating and gene flow were evolutionarily advantageous in recent time, might one also assume that moving genes between species has been advantageous throughout evolutionary history? The fossil record confirms that populations of *Homo erectus* were present across much of Asia during the Early and Middle Pleistocene. If dispersing Neanderthals or Denisovans had encountered any of these earlier hominins, they probably would have viewed them as potential mates. In fact, some researchers have raised the possibility that Denisovans are actually hybrids between eastward-dispersing Neanderthals and resident *Homo erectus*. Some support for this hypothesis comes from the fossil record. Two of the teeth found in Denisova Cave with Denisova-like DNA are extremely large for a Neanderthal—around 1.5 centimeters (just over half an inch) across. These archaic-looking teeth would have been useful for a diet that was different from that of later hominins—one that relied on grinding up tough food like grasses. The DNA data can also be interpreted as lending support to this hypothesis. Rather than encounter a southerly Denisova-like hominin that is absent from the fossil record, modern humans dispersing into Oceania may have encountered and hybridized with late-surviving *Homo erectus*, or perhaps an as yet undescribed lineage or hybrid lineage. This could explain why Denisovans (in this scenario Neanderthals with *Homo erectus* ancestry) and modern Oceanians (modern humans with both Neanderthal and Denisovan ancestry) have similar but not identical patterns of archaic ancestry compared to modern humans from other parts of the world.

This is, of course, conjecture, and based only on my and others' interpretation of the data available today. Inevitably, the next recovered fossil or sequenced ancient genome will have us erasing and rewriting human history once again. The only thing that we can know for certain is that we don't yet understand the details.

MEANWHILE, IN AFRICA

Simultaneously with the rise and spread of Neanderthals and Denisovans in Eurasia, several *Homo* lineages were evolving in Africa.

Based on data available today, our own lineage, Homo sapiens, probably evolved prior to 350,000 years ago and, for at least the first hundred thousand years or so after that, at least two Homo lineages coexisted in Africa: us and Homo naledi, a smaller-bodied and smaller-brained hominin that was discovered in the Rising Star cave system in South Africa in 2015. Precisely where Homo sapiens first emerged in Africa remains unknown, but by 100,000–200,000 years ago, fossils belonging to our lineage are found throughout the continent. Genomic data from living people suggest that these very early human populations were small and mostly isolated from one another, but that DNA flowed between them intermittently as populations expanded and contracted along with shifting habitats.

Something else was going on, too. By around 300,000 years ago, during the Middle Stone Age, archaeological sites across Africa begin to reveal evidence of an increasingly complex set of behaviors. At Jebel Irhoud, the early modern human archaeological site in Morocco that dates to 315,000 years ago, people were roasting stones to make them easier to flake and shape into tools. By roughly 100,000 years ago, Homo sapiens in northern, eastern, and southern Africa were making beads from shells and sometimes ostrich eggs and decorating these ornaments using pigments, sometimes with sophisticated geometric designs. Other late Middle Stone Age innovations by early Homo sapiens include fishing, trapping small animals, long-distance transport of materials like obsidian, developing mortuary rituals, and creating composite tools including projectile weapon systems. These are all evidence of increasing technological, behavioral, and cultural complexity.

How did this behavioral complexity—sometimes called behavioral modernity—arise, and how quickly did it happen? Until recently, many paleoanthropologists believed that what is now thought of as modern human behaviors arose suddenly, perhaps with a single genetic change. More complete and older archaeological records from across the African continent are rewriting this story. Today, most people working in this field believe that behavioral modernity evolved gradually over hundreds of thousands if not millions of years as innovations led to cultural and technological changes that led to more innovations. The archaeological record does seem to show a sudden increase in the rate of technological advance within the last 50,000 to 100,000 years, and scientists are exploring what roles things like growing populations, long-distance movements and consequent cultural exchanges, and even genetics may have played in our final transformation into behaviorally complex modern humans.

DNA from Neanderthals and Denisovans may once again provide some clues, or at least instruct us where to look in our own genomes for answers. We know that most living humans trace somewhere between 1 percent and 5 percent of our ancestry back to admixture with our archaic cousins. We don't all have the same bits of archaic DNA, however. Instead, different people inherited different stretches of archaic DNA. In fact, if we were to piece together all the bits of the archaic genomes that are circulating in people today, we could put together almost 93 percent of the Neanderthal and Denisovan genomes.

And the other 7 percent? This is where it gets interesting.

THE GENES THAT MAKE US HUMAN

A child with one Neanderthal parent and one modern human parent would be born with one complete copy of each parental genome. In that child's sperm or egg cells (depending on whether they are male or female), those genomes will break and then "recombine" approximately once per chromosome, creating new chromosomes that are a combination of the child's two genetic ancestries. Each sperm or egg cell will contain a genome that is 50 percent Neanderthal and 50 percent human. If that child mates with a modern human, their offspring will have one copy of the genome that is 50 percent Neanderthal and 50 percent human (the copy from their hybrid parent), and another copy that is 100 percent human. Those genomes will recombine, creating in that child sperm or eggs that have, on average, about 25 percent of the Neanderthal genome. Assuming there is no additional input of Neanderthal DNA, this dilution will continue through the generations. Today, many of us have a small amount of archaic DNA in our genomes, and chances are good that we've inherited archaic DNA from both of our parents.

When our archaic and human ancestors met and exchanged DNA, their lineages had been evolving along separate evolutionary paths for hundreds of thousands of years. Over time, mutations arose in stretches of DNA, and some of these were important in making each lineage unique. When they bred and their genomes recombined, some babies would be born that were missing important lineage-specific mutations. If a baby born to a human mother was missing an important human-specific mutation—having inherited that part of his genome instead from his archaic parent—that baby might die, or might not be able to thrive in a population of behaviorally complex humans. Over time, those stretches of DNA that didn't "work"—where a human could not survive without the human version of that stretch of DNA—would be tagged by natural selection and kicked out of the human gene pool. These stretches of DNA will be among the 7 percent of the archaic genome that doesn't exist in humans today. And this is where in the genome we should look to discover what it is that makes us different.

Today, tens of thousands of human genomes and a handful of high-quality archaic genomes have been sequenced. These make it possible to compile a list of the stretches of the human genome where no, or at least very few, living humans have inherited archaic DNA: the crucial 7 percent. The next step is to sort this list into stretches of DNA that were lost by chance and stretches of DNA that were rejected because they were incompatible with being human. This next step is hard, in particular because scientists still don't fully understand what all the parts of the genome are actually doing. We know how to find genes, and how to recognize parts of the genome that are controlling when genes are turned on and off. But we are still learning about the importance of things like interactions between genes, spacing between genes, and other elements of the genome that probably provide important but uncharacterized functions.

In my and other research groups, we've begun by looking at the parts of the genome that science best understands: the genes. Within a gene, some mutations are more impactful than others. A mutation that changes the sequence of a transcribed protein, for example, is more likely to make a functional difference than one that doesn't change the protein. We can measure how impactful a mutation has been by estimating how common it is among living humans. If all or most humans share a mutation in a stretch of DNA where nobody has the archaic version, the chance that this mutation somehow benefited early humans is high.

Recently, Ed Green, Nathan Schaefer, and I used this approach—identifying stretches of DNA where no human has archaic DNA and where all or most humans share a mutation that evolved after humans split from our archaic cousins—to identify what we call the human-specific genome. We found that the human-specific genome amounts to just 1.5 percent of our DNA. Not 7 percent, but much, much less. Now, we and others are starting to look more deeply into the genes in this 1.5 percent of the genome, searching for clues that will help us to understand what makes us human.

One of the genes in the human-specific part of our genome is Neuro-Oncological Ventral Antigen 1, or *NOVA1*. *NOVA1* is called a master regulator because it controls how fragments of genes are spliced together to make different proteins. Intriguingly, *NOVA1* is mostly active during early brain development, and people born with new mutations in their *NOVA1* gene often develop neurological disorders.

All living humans have a version of *NOVA1* that is different from the version found in all other vertebrates, including Neanderthals and Denisovans. The difference is small: Our version contains a single mutation. That all humans share this mutation, however, is good evidence that our version of NOVA1 behaves differently than that of our archaic cousins. To figure out what this mutation actually does, Ed Green and I collaborated with Alysson Muotri's lab at University of California, San Diego. Cleber Trujillo, a postdoctoral scholar working with Alysson, edited human cells so that their genomes included the archaic version of NOVA1, and then transformed these cells into brain-like organoids growing in dishes in his lab. As the brain organoids grew, Cleber tracked changes in cell size, shape, and mobility and sent us data so that we could analyze what proteins were produced. Cleber saw that the brain organoids with the archaic version of NOVA1 grew more slowly than those that had not been edited. The organoids with the archaic version had unusual, almost puffy-looking surfaces compared to the much smoother surfaces of organoids developed from cells with the human version of NOVA1. When Edward Rice and Nathan Schaefer, students in Ed's and my labs, analyzed the data, they saw that hundreds of genes appeared to have been spliced together differently depending on which version of NOVA1 the organoids had. Many of the differently spliced genes are involved with crucial functions during brain development, such as neural cell growth and proliferation and the formation of connections between synapses. Although certainly exciting, this is only the first experiment to have been completed and where the story ends for now. Future research, including our own, will have to figure out how common these differences are in different human cell lines, and what all these differences actually mean for human physical and cognitive development. This isn't the answer to what makes us human, but it does point us in a promising direction.

The 1.5 percent of our genomes that is uniquely ours probably contains many interesting and important clues to what makes us different from our archaic cousins. Lots of the genes in the human-specific regions of the genome are involved in some way with brain development. Others affect our diet and digestion, immune system, circadian clock, and dozens of other critical processes. But let's return briefly to the 93 percent of our genome where humans can just as readily inherit DNA from our archaic cousins as from other humans. There are two important points to make about this part of our genome.

First, it has become evident by looking at patterns of inherited DNA in populations around the world that it sometimes benefited humans to inherit the archaic version of a particular gene. People who live today in high-altitude Tibet, for example, are much more likely than people who live at low altitudes to have a version of a gene called *endothelial PAS domain protein 1*, or *EPAS1*, that evolved in archaic people related to the Denisovans. The archaic version of *EPAS1* alters production of red blood cells in a way that benefits people living in environments where there is less oxygen in the air. This means that ancestors of living Tibetans who inherited the archaic version of *EPAS1* were better able to survive in their high-altitude habitat than those who inherited the human version.

EPAS1 is far from the only example in which archaic DNA appears to have benefited modern humans. Several modern human populations have high frequencies of archaic variants of genes associated with immunity, presumably because the archaic version of these genes conferred an advantage to those who inherited them upon exposure to local pathogens. Archaic versions of genes associated with metabolism have also become common in some human populations, as have archaic versions of genes associated with skin and hair pigmentation. The high frequency of blue eyes in some European populations, for example, has been attributed to DNA that moved into the human population from Neanderthals.

Second, it is a little bit astonishing that so much of our archaic cousins' genomes persist today, including mutations that evolved in Neanderthals and Denisovans as they adapted to their habitats. Paleontologists often speak of Neanderthals as one of the first species that humans drove to extinction after evolving into behaviorally complex *Homo sapiens*. Our genomes, however, reveal this to be an oversimplification. Our ancestors didn't simply outcompete the Neanderthals, they also used them to improve themselves. The disappearance of the Neanderthals was a harbinger of things to come.

— CHAPTER 3 —

Blitzkrieg

D URING THE SUMMER OF 2007, WHILE VISITING AN EXTRA-
ordinary museum in Moscow, Russia, I did something regret-
table. Ignoring the protestation of my colleagues, I reached out and
picked up a 50,000-year-old fossil: an isolated horn that once belonged to a Siberian woolly rhinoceros. In that moment, I was filled
with awe. I was holding an intimate part of a long-dead animal—
one of the last of its kind, a descendant of tens of millions of years
of evolutionary innovation, a cousin of some of the most threatened
species on the planet today. I understood that this rhino's death, and
indeed the death of its entire species, may have been our species'
fault. And here was his horn, on a bottom shelf in an overcrowded
museum, surrounded by animated reassembled cave lion skeletons
and gaudy contemporary ivory carvings for sale, all basking under
an artificial sunset. The setting was both perfectly fitting and entirely unfair, a coarse reminder of the biological diversity encountered by our ancestors juxtaposed against our greediness. More than
a decade later, the lingering stench of my decision to pick up that
horn still conjures an uncomfortable mixture of embarrassment
and remorse.

The visit to the museum was not meant to be dispiriting. I was in Moscow with a small group of scientists with interests in different aspects of mammoth evolution. We were heading to the Fourth International Mammoth Conference, which was to convene a week later in Yakutsk. First, though, our small coalition from Africa, North America, and Europe would spend a week as tourists and guests of Andrei Sher, a Russian paleontologist who spent much of his career studying ice age animals, and his colleague, entrepreneur and mammoth enthusiast Fedor Shidlovskiy.

Our pre-conference week in Moscow was a whirlwind of unpredictable adventures. On our first day as tourists, we were stopped by the police for driving on the sidewalk while threading through traffic to visit Saint Basil's Cathedral. The next day, as we were driving through a pedestrianized park (clearly not having learned our lesson), we passed an elephant ambling by a replica of the Vostok rocket. When we relayed this experience to Hezy Shoshani, an elephant expert and conservationist who was part of our group but not with us on that particular journey, he expressed neither shock nor urgency but instead countered with a sober "African or Asian?" We were treated to extravagant dinners in the Moscovian suburbs and taken shopping on Arbat Street. But the real reason for our week in Moscow was to peruse the collections in Shidlovskiy's museum for specimens that might help us in our research.

Shidlovskiy's Ice Age Museum was the physical manifestation of his passion for the dead creatures of the Siberian tundra. It was housed within the All-Russian Exhibition Centre, an expansive city park filled with once-extravagant exhibit halls, garish fountains featuring gold- and bronze-plated statues, and replicas of Soviet-era technological accomplishments. The entrance to the Ice Age Museum was almost invisible among the aging buildings, whose tenants by the early 2000s were mostly small shops selling anything from pre-owned clothing to machine guns to hand-painted matryoshka dolls. The door to the museum opened at the base of a narrow, steep flight of stairs whose precarious ascent was further impeded by the strategic placement of a large cardboard box filled with blue paper shoe covers. A handwritten notice taped to the wall above the box declared that entrance would be forbidden to those with inappropriately dressed shoes, and so we crowded into the stairwell and fumbled awkward attempts to dress each shoe in a disposable paper bag without jamming our elbows into the walls or one another. Once our shoes were appropriately attired, we climbed the stairs into the main hall of the museum, where we were in for many surprises.

Before it closed in 2018, the Ice Age Museum, which billed itself as a "museum-theatre," was a popular spot for school tours and families with young kids looking to get a taste of the ice age. When I was there in 2007, visitors emerged from the entry into a hall filled with stuffed life-sized versions of extinct species onto which people were encouraged to climb. Moving deeper into the museum, rows of skulls and large bones led to displays of fully articulated skeletons of steppe bison, cave lions, and mammoths, made piecemeal from bones unearthed in Siberia by Shidlovskiy and his friends. The central hall featured a mechanical baby mammoth desperately and perpetually trying to get out of a pit into which it had fallen, surrounded by screens looping videos of Shidlovskiy on his Siberian expeditions. In the back hall, a life-sized, lavishly ornate throne made entirely of carved ivory was on display and for sale, as were various pieces of ivory scrimshaw, ivory tables, ivory chess sets, ivory statues, and ivory jewelry. Shidlovskiy's assistant assured us that it was all mammoth ivory.

It was in the context of this sensory deluge that I made my tactical blunder. Along one of the museum's walls was a row of particularly unusual specimens: an enormous mammoth femur, a perfectly intact cave lion skull, several massive bright yellow-white mammoth tusks, lines of mammoth and mastodon teeth of all sizes, and a piece of rhinoceros horn. I had never seen a rhino horn up close, as I'd done most of my museum work to this point in North

America, where woolly rhinos, as far as we know, never were. The horn was gorgeous. It was dark brownish gray and much larger than I'd expected, though it was the smaller of a woolly rhino's two horns. When I picked it up, I felt that it was also heavier than I imagined it would be. Also rough, and bumpy. Not at all like the hundreds of bones that I'd handled from ice age animals.

I stared into the horn, transfixed. I thought about the first living rhino that I'd seen, a tame black rhino named Morani who lived at Ol Pejeta Conservancy in Kenya. I remembered the guards who protected Morani laughing at me as I trembled in fear while leaning on the sleeping giant for the requisite photo. I thought about the rhinos that had died in Ol Pej and elsewhere since then, most of them murdered by poachers who sold the horns on the black market to swindlers who turned them into fake elixirs to sell to other desperate people. I thought about the woolly rhinos that survived the peak of the last ice age and, at the precipice of extinction, found themselves at the mercy of sophisticated human predators they were unable to outrun.

I looked up at my friends, expecting them to be experiencing the intensity of the moment with me, or at least to be waiting anxiously for their turn to hold the horn. That is when I noticed the expressions on their faces: fear, disgust, horror, amusement. I smiled uncomfortably and held the horn out to Ian Barnes, another ancient DNA researcher with whom I'd worked for years. Ian threw his hands up and inched backward, sliding his paper booties across the wooden floor as he shook his head. That was when my ears finally tuned in to their words: "Don't touch that!" "Why would you do this?" "Poor choice, Shapiro." "There's not enough soap in Moscow to fix this." Confused, I put the beautiful fossil back on the shelf and, making my second worst decision that day, wiped my hands on the sides of my pants. They stared. Ian chuckled. I looked behind me, hoping they might be reacting to something else. There was nothing behind me. Feeling self-conscious, I glared confusedly at Ian and raised my arms into an aggressive shrug. As the upward motion of my hands stirred a burst of air toward my nose, I smelled it.

Woolly rhinoceros horn, as I was reminded in that moment, is made of hair. Tightly compressed threads of keratin. Over time, as the threads break down, it rots. Let's just say that 50,000-year-old dreadlocks, never-washed, buried for a while, dug up, and placed on a shelf in a warm room for another while, are not the kind of thing that you want to spend any time handling with bare hands.

As I headed to the bathroom to find the first of the several bars of soap that I would exhaust that day, I couldn't help but smile. I had, after all, just held part of an extinct woolly rhinoceros.

THE SIXTH MASS EXTINCTION

Rhinoceroses have been around for a long time. Over at least the last 50 million years, some 250 different rhino species evolved and then went extinct. Some rhinos were small and fat and looked a bit like squat horses. The largest land mammal that ever lived was a type of rhino. Rhinos lived in the tropics, in temperate zones, at high altitudes, and in the Arctic. Some lived on land, and others immersed themselves in rivers where they filled the niche of modern hippos. Some rhinos had tusks that stuck out from their lower jaw, and others were both tuskless and hornless. But the most famous rhinos had horns: sometimes only one horn that curved upward from the top of their nose; sometimes two horns, either poking out side by side or stacked one in front of the other. The number and diversity of rhino species have been declining since the Early Miocene some 23 million years ago, but living rhinos are no less awe-inspiring than their ancestors were.

Five rhino species are alive today, although most are just hanging on. The Sumatran rhino, *Dicerorhinus sumatrensis*, lives in the tropical and subtropical forests of Indonesia and Malaysia, where they declined from 800 or so individuals in the mid-1980s to fewer than 100 today. The Javan rhino, *Rhinoceros sondaicus*, exists as a single population of around 60 individuals in western Java, Indonesia. The Indian or greater one-horned rhino, *Rhinoceros unicornis*, was declining but is now a conservation success story. Found throughout India and Nepal, the Indian rhino has recovered from fewer than 200 individuals in the early twentieth century to around 3,500 today. Two rhino species live in Africa: the black rhino, *Diceros bicornis*, and the white rhino, *Ceratotherium simum*. In 1970, an estimated 65,000 black rhinos lived in southern and eastern Africa. Unfortunately, demand for rhino horn for fake medicines reduced this population by 96 percent by the turn of the twenty-first century. Thanks to protection from poaching, however, this population is making a slow comeback, with nearly 5,000 individuals alive today. White rhinos have a divided story. Of the two subspecies, the southern white rhino is doing well compared to other rhino species, with around 20,000 individuals estimated to live in the savannahs of southern Africa. The northern white rhino, though, is functionally extinct. Only two individuals remain, Najin and her daughter Fatu, and they live under 24-hour protection at Ol Pejeta Nature Conservancy in Kenya. Sudan, who was Najin's father and Fatu's grandfather, died in March 2018. He was 45 years old.

The extinction of the northern white rhino will be the first rhino extinction in around 14,000 years, when the last of the woolly rhinoceroses, *Coelodonta antiquitatis*, disappeared from Siberia's steppe tundra. Woolly rhinos get their name from their thick fur coat, which kept them warm in the Far North during the coldest phases of the Pleistocene ice ages. A second ice age rhinoceros, the Merck's rhino, *Stephanorhinus kirchbergensis*, preferred slightly warmer habitats than woolly rhinos and disappeared earlier, probably before 50,000 years ago. A third lineage, *Elasmotherium sibricum* or the Siberian unicorn (named after their single, long horn), lived in the cold herb-dominated grasslands of the central Asian steppe and went extinct around 36,000 years ago.

Why did these three cold-adapted rhinos go extinct? As with the extinctions of other iconic ice age animals, the debate has focused chiefly on two potential causes: climate change and humans. Teasing apart their relative culpability has been a mainstay of paleontological and archaeological research for decades, and, more recently, for researchers working in ancient DNA.

The case against humans is strong. Neanderthals, Denisovans, and modern humans were all living in Eurasia by 50,000 years ago, where they were hunting and eating megafauna. By 36,000 years ago, Siberian unicorns had gone extinct and woolly rhinos had disappeared from Europe. Neanderthals were also gone, and modern humans were increasing in abundance across woolly rhino habitats. Woolly rhinos were extinct by 14,000 years ago, and by then modern humans were everywhere; they had even spread across Siberia and the Bering Land Bridge into the New World. The coincidence in timing between the rhino extinctions and the increase in human abundance is undeniable. But were humans actively hunting woolly rhinos? In Siberia, the archaeological evidence suggests that people were consuming rhinos but not often: woolly rhino remains are found at 11 percent of archaeological sites younger than 20,000 years old, but rhinos are never the only prey found at a site. It is impossible to know whether this was because woolly rhinos were rarely used as prey or because woolly rhinos, whose range was shrinking, were simply rare.

The case against climate is also strong. Around 35,000 years ago, the Eurasian climate transitioned into an interstadial period—a cold period that was not quite cold enough to qualify as a glacial period. Summers became cooler, winters became more erratic, and nutrient-rich grasslands were replaced by mosses, lichens, and other tundra plants. Signals from carbon and nitrogen isotopes in the bones of the animals living on the tundra tell us that saiga antelopes, which shared the steppe landscape with Siberian unicorns, shifted their diet to eat these tundra plants. Siberian unicorns, however, did not shift their diets. Perhaps they couldn't—their teeth, large horn, and low-slung head suggest an animal exquisitely adapted to feeding on grasses that grow close to the ground. The climate continued to cool, and by 20,000 years ago had plunged into the coldest part of the last ice age. The woolly rhino, which was able to feed on some mosses and lichens in addition to grass, was the only rhino still alive in the Far North. Their range, however, had been reduced to whatever isolated patches of grassland remained across eastern Siberia. The last woolly rhinos lived around 14,000 years ago, which, coincidentally, is when the climate went through a major warming phase. The grassland habitats on which the woolly rhinos relied were replaced by shrubs and trees.

Which was it? Are humans to blame for the extinctions of the cold-adapted rhinoceroses, or is climate change to blame? For now, the favored hypothesis is that both are to blame. The transition into the interstadial 35,000 years ago probably reduced rhino habitat just as people—Neanderthals or modern humans or both—suppressed their populations through hunting to the point that they became vulnerable to any change to their habitat. In northeastern Siberia, the replacement of grasslands by shrubs and trees after the peak of the last ice age also coincided with increasing hunting pressure by humans. In the spirit of innocent until proven guilty, the evidence from the Eurasian fossil and archaeological records is, as of today, insufficient to convict us for the crime of rhino extinction. But the coincidence in timing is conspicuous. And, as we will see, it only scratches the surface as far as coincidental timing is concerned.

The last 50,000 years have seen an undeniable surge in the rate and number of species extinctions. Although published estimates vary, most scientists agree that the rate of species extinction today is more than 20 times higher than the background extinction rate—the normal rate of extinction in geological history. We are living in the midst of a mass extinction event, the sixth in Earth's history. What began with the loss of the megafauna—woolly rhinos and mammoths, for example—continues today with the losses of microfauna—snails and bees, for example—as well as fishes, songbirds, wildflowers, and trees. These extinctions have cascading effects. Extinction disrupts food webs, dismantles ecological interactions, and denudes landscapes.

That some recent extinctions are our fault is undeniable. People hunted California golden bears to extinction during the first quarter of the twentieth century, converted the entire habitat of Caspian tigers to cropland by the 1970s, and caught in illegal gill nets nearly all remaining Sea of Cortez vaquitas by 2020. The consequences of these near-time extinctions ripple through the food webs in our own backyards. When large carnivores disappear, for example, the large herbivores on which they prey expand and overeat grasses, trees, and shrubs, which reduces habitat for smaller herbivores and causes their populations to crash, which in turn causes the smaller carnivores that prey on them to decline, and on and on and on. From these near-time extinctions, it is clear that our actions, not only alter the evolutionary trajectories of the species that go extinct (as an understatement), but also fundamentally change the evolutionary landscape in which other species, including our own, live.

THE FIRST VICTIMS

In 2017, a new dating analysis of stone tools excavated from Madjedbebe rock shelter in northern Australia shocked the paleoanthropological world by proving that humans were there 65,000 years ago. This date pushed back the previously assumed date of human arrival in Australia by as much as 15,000 years. If the new dates are correct, these early Australians were either on the leading edge of a rapidly moving band of humans already capable of purposeful ocean voyaging, or part of an earlier exodus from Africa. Only a decade ago, a scientist arguing in favor of two separate African exoduses might have been laughed off the presentation stage, but evidence is accumulating from both genomics and archaeology that supports this idea: 200,000-year-old Neanderthal fossils from Germany with DNA evidence that their ancestors already had intimate liaisons with *Homo sapiens*, a 180,000-year-old *Homo sapiens* jawbone from a cave in Israel, and *Homo sapiens* teeth and other skeletal remains dating to between 70,000 and 125,000 years ago from four caves in China. If these fossils, dates, and ancient DNA data are indeed evidence of an earlier wave of modern humans dispersing out of Africa, then it is possible that some of these people made it to Sahul, and then to Madjedbebe by 65,000 years ago.

But then what happened? As of yet, no other archaeological site in Australia dates to within 10,000 years of Madjedbebe, and no DNA evidence has been recovered from these earliest Australians. The people of Madjedbebe may have been part of an early wave of dispersing humans that either went extinct or was replaced by later people, or they could have been the front-runners of the single dispersing wave. By as early as 55,000 years ago, however, and certainly by 47,000 years ago, people had colonized much of the Australian continent, and have been there ever since.

The first Australians arrived to find a continent that had been drying out over the preceding several glacial cycles. The widespread, dense, fire-prone eucalyptus forests of the Early and Middle Pleistocene had, by around 70,000 years ago, been replaced by less flammable open shrubland, which was populated by a magnificent diversity of indigenous fauna. People spreading across the continent would have encountered giant wombats—grazing herbivores the largest of which weighed in excess of 2,700 kilogram (6,000 lb). They would have met giant kangaroos, echidnas the size of present-day sheep, and marsupial lions. They would have seen enormous snakes and crocodiles and absurdly tall flightless birds—"daemon ducks of doom"—that weighed more than 700 kilogram (1,600 lb). By 46,000 years ago, all these animals were extinct.

Is the coincidence in timing between the disappearances of the Australian megafauna and the colonization of Australia by humans just that: a coincidence? Australia experienced pulses of widespread fire and a slow deterioration of the climate over the hundreds of thousands of years prior to human arrival. The fossil record, however, does not show commensurate declines in Australian megafaunal populations during this prehuman interval. Instead, Australian megafauna decline abruptly at roughly the same time that humans spread across the continent. No continental-scale sudden changes are recorded in the Australian climate around the time that humans first appeared. Local climates were shifting but, because the megafauna belonged to taxonomically diverse groups, ate a variety of plants and prey, and could survive in different habitats, they had opportunities to move to or survive in refugia. But they did not.

The antiquity of these events in Australia—shifting vegetation and the changing fire regime, human arrival, and megafaunal extinctions—has made it difficult to pin down their precise order and timing. During the second half of the twentieth century, radiocarbon evidence accumulated that some Australian megafauna survived until as recently as 30,000 years ago. These dates were used as evidence that humans coexisted with the Australian megafauna for more than 15,000 years and therefore could not be responsible for their extinction. Re-dating these remains using new approaches, however, has revealed that all these fossils are actually much older. Humans were widespread in Australia by at least 47,000 years ago, and the Australian megafauna were extinct by around 46,000 years ago, give or take a few thousand years depending on region and measuring error. Humans coexisted with the indigenous megafauna of Australia, but not for long.

Some of the most pungent evidence of human culpability in the Australian megafaunal extinctions comes from megafauna poop. A few years ago, a team of scientists led by Sander van der Kaars from

Monash University in Victoria, Australia, and Giff Miller, from the Institute of Arctic and Alpine Research and the University of Colorado in the United States, took a new approach to reconstructing the Pleistocene history of Australia. They cruised just offshore from southwestern Australia, stuck a giant coring machine into the ocean floor, and sucked up a long stick of mud. This stick of mud was made up of layer after layer of dirt, pollen, DNA, and other organismal remains that had been blown into the ocean from the continent and settled onto the ocean floor. They examined these layers from the bottom (oldest) to the top (youngest) layers of the core, collecting data that they used to reconstruct a timeline of habitat changes on the nearby continent.

The team determined from a combination of data from radiocarbon and other chemical signatures that their stick of mud contained the last 150,000 years of history of the flora and fauna of southwestern Australia. They learned from pollen preserved in the core that the forests of southwestern Australia were warm and wet around 125,000 years ago, during the last warm interglacial period, and transitioned to dry-adapted vegetation at the onset of a glacial period 70,000 years ago. The rate of sediment accumulation from 70,000 until 20,000 years ago revealed that the climate during that entire interval was very dry and cool. Layers of charcoal found within this dry interval told them both when large fires happened and what type of vegetation was burning: 70,000 years ago marked a shift from frequent and intense eucalyptus fires to less intense and less frequent vegetation fires. And they found a lot of poop. Or, more accurately, the fungal evidence of poop, *Sporormiella*.

Sporormiella is a genus of fungi that are obligately coprophilous, which means they grow exclusively on poop. Unlike the poop itself, spores from *Sporormiella* are hardy and accumulate readily in sedimentary records. *Sporormiella* are so common and so well associated with herbivores that they are often used as a proxy for them in paleoecological studies: where we find *Sporormiella*, we can assume that megafauna are present. And when *Sporormiella* disappears, we can conclude that so, too, have the megafauna.

A few years ago, I was part of a research team that used *Sporormiella* to figure out when and why mammoths went extinct on St. Paul Island, a tiny island in the Bering Sea west of the Alaskan mainland. St. Paul was cut off from the mainland by sea level rise around 13,500 years ago. Despite their isolation from the mainland population, mammoths survived on St. Paul for nearly 8,000 years after it became an island. Mammoths had no predators or competitors on St. Paul; they were in fact the only terrestrial large mammal on the island and humans did not arrive there until a few hundred years ago. The cause of their extinction has therefore been a mystery.

To solve this mystery, we extracted a sediment core from the bottom of the only freshwater lake on St. Paul—an old volcanic caldera—and then, as the Australian team did with its ocean core, examined the layers of our core from bottom to top. We looked for pollen and plant macrofossils that might tell us if the vegetation changed and mammoths ran out of food. We identified small insects and crustaceans that could tell us if the water was turbid or salty. We looked for mammoth DNA directly, as mammoths would have waded into the lake to drink the fresh water, depositing their DNA in the process. And we looked for *Sporormiella*.

We found lots of mammoth DNA and *Sporormiella* spores from the bottom of the core and all the way up the core until around 5,600 years ago, when we suddenly found none of either. That meant that large herbivores were there—and that meant mammoths, which were the only large herbivores on St. Paul Island—until 5,600 years ago, after which they were not. We found no changes in the pollen at that time interval, which ruled out a change to the plant community that might have caused mammoths to starve. However, we found other changes. The rate of sedimentation increased and, as the lake got shallower, it also got saltier. The insects and crustaceans that we found in the core changed from species that thrive in deep,

clear, fresh water to species that can tolerate lots of floating par-
ticulates. Together, these data provided our answer. Around 5,600
years ago, the lake, which was the only source of fresh water on St.
Paul, nearly dried up. Mammoths went extinct because of an intense
drought.

The team working in Australia did not expect to see a disappearance of *Sporormiella* from its cores, as herbivorous mammals still
live on the continent. However, by counting the number of spores
along the core, they could infer changes in herbivore abundance
over time, which would reveal when herbivore populations were
large and when they were small. The team found that *Sporormiella*
made up about 10 percent of the total pollen and spore count in the
marine soils starting at the bottom of the core. Then, around 45,000
years ago, the amount of *Sporormiella* in the core declined precipi-
tously, bottoming out at around 2 percent of the total pollen and
spore count by 43,000 years ago. Their data showed that many fewer
herbivores were pooping in the southwestern Australian forests after
43,000 years ago, when first humans arrived in the region.

Is the case against us closed? Maybe. The evidence of our culpability is mostly coincidental. There is some archaeological evidence
that humans were hunting and consuming Australian megafauna.
Giant wombat bones have been recovered at several early archaeological sites but, so far, none have been found with cut marks made
by humans. The most compelling evidence of early Australians
hunting megafauna comes from fragments of burned eggshell from
the extinct large waterfowl *Genyornis* that have been found at a
few sites across the continent. Because of their size, even low rates
of hunting could have had disproportionate effects on Australia's
megafauna. Larger animals tend to have fewer offspring and slower-
growing populations than smaller animals, which means that the
same amount of hunting pressure—even as little as one person
taking one juvenile per decade—will push a large animal toward
extinction faster than it would push a smaller animal toward ex-

tinction. This explanation has been dubbed by Australian ecologists Barry Brook and Christopher Johnson as an "imperceptible overkill" scenario, in which the impact of early Australians was sufficient to cause the extinctions though this impact was not recorded in the archaeological record.

Regardless of whether humans were directly or indirectly or not at all responsible for the extinction of the Australian megafauna, the disappearance of so many large herbivores had immediate and lasting impact on Australia's ecosystems. Large herbivores eat lots of plants, and this maintains the openness of forest and shrub ecosystems and removes fuel for fires. Large herbivores also disperse seeds, often over long distances, and recycle nutrients through digestion and by turning over the top layer of soil as they walk. When Australia's megafauna disappeared, the forest communities changed. Forests became drier and denser, and fires became more frequent, more widespread, and more intense. Plants and animals unable to tolerate the new fire regime or altered communities either moved or went extinct. Australia's ecosystems, and the environment in which Australia's plants and animals (including people) lived, were fundamentally altered.

NEW WORLD, NEW PREY

Sometime during the most recent ice age, a small group of people living in northeastern Asia ventured in a direction that no one had gone before. It was cold, very cold, and food may have been scarce, forcing small family groups to fan out in search of food. As they set out toward the east, the scenery would have been much the same: no tall trees, waning grasslands, and many, many mosquitoes. As they moved, these dispersing family groups may have followed a coastline and foraged from the oceans. Or perhaps they followed an inland route, tracking game like bison or mammoths. They could not, however, have been aware that the route they were taking was one that no other human had taken, or that the land on which they were walking would someday be tens of meters below sea level, after global warming melted continental glaciers and sea levels rose once again. They also could not have known that they were about to become discoverers of the New World, which, of course, they would not know as such.

Archaeological evidence of humans in Beringia is rare. The oldest known site, the Yana Rhinoceros Horn Site, is located at the western geographic extreme of Beringia in northeastern Siberia. Archaeologists working at the Yana site have found stone scrapers, tools, and spear shafts made from wolf bones, rhinoceros horn, and mammoth ivory, as well as butchered mammoth, muskoxen, bison, horse, lion, bear, and wolverine bones. Many of these bones and tools are around 30,000 years old. The earliest evidence of humans on the other side of the land bridge is from the Bluefish Caves site in Yukon, Canada, which is at Beringia's easternmost extreme. There, archaeologists have found human-modified bones of horses, bison, sheep, caribou, and wapiti, some of which date to 24,000 years ago. The Yana and Bluefish sites are both the oldest archaeological sites in Beringia and the only Beringian archaeological sites that date to around or just before the peak of the last ice age. With data only from these two sites, we cannot know exactly when people first dispersed onto and across the land bridge or how abundant these early inhabitants were. We do know, however, that people were there, somewhere, during the coldest part of the last ice age.

Life in Beringia during the ice age probably wasn't terrible. The climate was too dry for the region to become covered in glaciers, but enough rain and snow fell each year to support a rich steppe tundra ecosystem. Rare saber-toothed cats, lions, and bears would have been formidable foes, and decent habitat was probably sparse, making it uncommon to run into someone from a different familial group. But the people of ice age Beringia lived in a habitat with plenty of edible plants and a decent selection of prey, including mammoths, bison, horses, and caribou. A good life, as long as one could avoid being eaten.

While people appear to have colonized the entirety of Beringia by the coldest part of the last ice age, farther human progress to the east or south had to wait for the ice age to end. By the time people were living in Bluefish Caves, a 4,000-kilometer-wide (2,500-mi-wide) ice sheet stretched from the southern coast of present-day Alaska to the western coast of present-day Washington State and across the continent to the eastern coast. This ice sheet effectively blocked any human dispersal out of Beringia until the climate warmed and the ice began to melt. Archaeologists theorize from genetic data that people were "stuck" in Beringia by this ice barrier for more than 7,000 years before they were finally able to colonize the rest of the continent. And while it is clear that people did eventually leave Beringia, when, how many times, and by what route this happened has been one of the most enduring arguments in American archaeology.

One of the first available routes for southward dispersal was along the western coast, where people would have had access to a reasonably hospitable maritime climate and whatever resources they could take from the sea and coastal freshwater ecosystems. Another, more controversial dispersal route went through the central continent, where prey such as bison would have been more abundant. The proposed continental route followed the touchpoint of the two smaller ice sheets that coalesced during peak glacial times: the Laurentide Ice Sheet, which sat atop the central and eastern continent, and the Cordilleran Ice Sheet, which ran along the western coastal mountains. As the climate warmed, these two ice sheets slowly shrank away from each other, creating a north-south ice-free corridor that led from Beringia through present-day Alberta and western British Columbia and into the continental United States.

When I first started thinking about this question, it was well understood that people were widespread in both North and South

America by 13,000 years ago. Dozens of archaeological sites on both continents provide compelling evidence that humans were established by at least then and perhaps a few thousand years earlier. One of these sites, Paisley Caves in Oregon, includes several pieces of 14,000-year-old human poop, which were identified as such by my friend Tom Gilbert, who leads an ancient DNA research group at the Natural History Museum of Denmark and is famous for trying to get DNA from pretty much anything. While one is left to wonder why these people pooped in their—or perhaps their neighbors'—cave, the discovery of human poop in Paisley Caves confirms that humans were in Oregon, south of the ice sheets, by 14,000 years ago.

A firm minimum occupation date of 14,000 years ago meant that our team could resolve which of the two routes people used to get to the North American midcontinent by answering a simple question: Was the ice-free corridor usable by that time? To be clear, the corridor would have appeared as soon as the glaciers started to melt, which happened prior to 14,000 years ago. But there is little chance that the newly formed path between the glaciers would have been usable immediately. The trip on foot from Beringia would have been long and arduous. People who entered the corridor would have had no idea where they were going, how long it would take to get there, or even that they were moving into a corridor between two massive melting sheets of ice. For people to enter the region of the ice-free corridor, the plants and animals on which they relied would already have to be there.

With this question in mind, Duane Froese and I, along with people from our labs and several archaeologist and paleontologist colleagues, set out to determine when the region of the ice-free corridor was the kind of place into which people would have willingly wandered. This project was facilitated by a quirk in bison genetics that made them ideal for testing when the corridor became passable. While the ice sheets were merged and movement was cut off between Beringia and the rest of North America, the population of bison living south of the ice almost went extinct, probably due to a combination of the decline of grasslands and competition with other herbivores like mammoths and horses. While the southern bison population was near collapse, they lost nearly all their mitochondrial genetic diversity; only a single variant remained. When the southern bison population recovered after the ice age, all the many tens of thousands (and eventually millions) of southern bison had this one mitochondrial variant, making them easily distinguishable from the bison that had spent the ice age in Beringia. To learn when the corridor was passable, all we had to do was collect bison bones from the corridor, extract DNA, and determine whether they were the southern or northern type. The earliest date that we saw either the northern-type bison in the south or the southern-type bison in the north would reveal when the corridor was both habitable and passable by bison. Because bison require a similar habitat to that of humans, this would, by proxy, reveal when the corridor region was first habitable by humans.

After genetically typing dozens of bison bones, we concluded that the ice-free corridor opened like a slow zipper from both ends. By 13,500 years ago, northern bison had begun to move southward and southern bison had begun to move northward. By 13,000 years ago, both northern and southern bison were present in the central corridor region. And by 12,200 years ago, southern bison were present at the northern end of the corridor. This meant that the corridor was open and passable only by 13,000 years ago—much too late to have been the route taken by southward-dispersing humans. By corollary, humans must have taken a coastal route southward from Beringia, probably thousands of years before the corridor was open and viable.

Intriguingly, the only evidence of bison actually transiting the corridor was in the opposite direction of what we expected: a few bison moved through the corridor from the south to the north rather than from the north to the south. In retrospect, a preference for south-to-north dispersal makes ecological sense. After the ice retreated, the southern and central regions of the corridor were quickly colonized by grasslands, open woodland, and boreal parkland—productive habitats that could have supported a diverse community of herbivores. The northern corridor region, alternatively, was colonized by alpine and shrub tundra, with patches of dense spruce forests. These habitats provided much less nutrition and would have been tough for large mammals to traverse. Given these ecological data, it is not surprising that grassland specialists like bison moved into the corridor from the south.

Species that hunt these grassland specialists appear also to have followed their prey northward through the corridor. Today's Alaskan wolves, which prey on bison, descend from a population of wolves that dispersed northward after the ice sheets melted. And when humans finally did use the corridor, they, too, used it to disperse northward; fluted point technologies developed earlier by southern people to hunt and kill bison are found in Alaska dated some 500 years after the corridor became passable.

A clearer picture of human colonization of the Americas emerges after the ice recedes and evidence of humans becomes more common. By 13,000 years ago, humans were widespread across the Americas and had become skilled hunters of the indigenous big game. And there were consequences. As people moved from Beringia into continental North America and then into South America, the native megafauna began to go extinct. Based on thousands of radiocarbon dates from fossil megafaunal remains, the first extinctions happened in Beringia roughly 13,300–15,000 years ago. The wave of extinctions in south-of-the-ice North America occurred 12,900–13,200 years ago, and in South America 12,600–13,900 years ago.

The ranges of dates that I cite here are tight, especially for continental North America, where dozens of taxonomically and ecologically diverse species become extinct within only 400 years. The late Paul Martin, an earth scientist who spent most of his career at the University of Arizona, was the first to translate the radiocarbon record into an explicit theory about the cause of the megafaunal extinctions. Now known either as the "overkill hypothesis" or "blitzkrieg," as Martin dubbed it after the German word that translates literally to "lightning war," Martin's theory holds that humans, upon encountering prey that were naïve to us, either were already proficient big-game hunters or became so opportunistically, driving down the numbers of our new prey in a manner that outpaced their capacity to reproduce. Paul Martin's blitzkrieg theory holds that these extinctions, and indeed megafaunal extinctions that occurred across the globe and coincidentally with human arrival, all had a common cause—us—and that the proof was in the radiocarbon record.

In the New World, though, paleontologists again face the tricky challenge of disentangling the roles of climate and humans in the extinctions. Unlike in Australia, where the climate was not changing when the megafauna began to disappear, the North American extinctions coincided with major climate changes and habitat rearrangements. The peak cold period of the last ice age ended around 19,000 years ago, but the climate remained cool for several millennia. Then, around 14,700 years ago, the climate transitioned suddenly into a warm and moist interstadial. This warm period lasted for several millennia and then shifted once again, this time into a cold period called the Younger Dryas. This second shift was particularly abrupt, with climate returning to glacial-like conditions within as little as a decade. The Younger Dryas was a period of increased seasonality—colder winters and hotter summers—which meant shorter growing seasons. Less food was available for herbivores, and what food was available was less nutrient rich, thanks to a decrease in atmospheric carbon. Then, around 11,700 years ago, the climate shifted for a third time, this time warming abruptly and marking the onset of the current Holocene warm period. As the climate oscillated between extremes, the consequent changes in temperature and precipitation patterns most impacted larger-bodied and slower-reproducing mammals, which are precisely those that became extinct.

I've spent much of my career trying to tease apart the relative roles of climate change and the spread of humans across North America in causing the extinction of the Beringian megafauna. My work and that of my colleagues has provided clues into how large mammals were impacted by these stressors. We've seen that not all species respond simultaneously to changes to their habitat, or even in the same way. In Eurasia, for example, muskox and woolly rhino populations grew and declined along with the availability of their habitat, but the success and failure of horse and bison populations was less closely linked to clear shifts in the global climate. Of course, global climate is probably a poor predictor of many species' success or failure. Widespread species are not likely to become suddenly extinct across their entire range, at least without a disaster like the asteroid that ended the reign of the dinosaurs. As ancient DNA data has become easier and less expensive to generate, we have begun to generate data from geographically isolated populations of the same species. These data are helping to disentangle the roles of local climate changes and humans in species' declines.

One of the first species to be studied in this way was mammoths. Ancient DNA isolated from mammoth remains from across the Northern Hemisphere has revealed that, rather than becoming suddenly extinct, mammoths declined slowly throughout the last 50,000 years. Over that time, geographically distinct populations went extinct at different times. In North America, for example, mammoths went extinct in the central continent during the Younger Dryas but survived in the far north of Alaska until around 10,500 years ago. But this was not the end of mammoths. Two small populations survived on islands for much longer: the St. Paul population until around 5,600 years ago and another on Wrangel Island, off the northeastern tip of Siberia, that survived until around 4,000 years ago.

The slow decline of mammoths across the Northern Hemisphere is not predicted by the blitzkrieg model, but neither does it mean that humans are off the hook for their disappearance. On Wrangel Island, for example, the timing of mammoths' final demise coincides with human settlement. Genetic data from some of the last of the Wrangel mammoths show, however, that the population was already in trouble when humans arrived, as many generations of inbreeding riddled their genomes with mutations that reduced their fitness. These data suggest that this last population of mammoths probably would have died out even if humans had never made it to Wrangel Island. But if our ancestors drove mainland mammoths to extinction, that ended the possibility that immigrant mammoths from the mainland might enrich the Wrangel gene pool and save this last population. Did humans indirectly doom the Wrangel Island mammoths to their genetic collapse?

Perched on my fence of indecisiveness, I see the story of the North American extinctions as, well, complicated. The climate was changing markedly during the transition into the Holocene. Then humans arrived and became the proverbial straw that broke the mammoth's back. If the megafauna were already in trouble, then human populations need not have been huge to have a catastrophic impact.

We're working in my lab on a large project that aims to learn how troubled the North American megafauna already were when humans started to spread across the continent. We have extracted ancient DNA from hundreds of bones from bison, mammoths, and horses that lived over the last 40,000 years in the region that was once eastern Beringia. We collected the remains of beetles that can be used as indicators of temperature and precipitation regimes. We counted ground squirrel nests and identified the plants they are made of, measured carbon, oxygen, and nitrogen isotopes from bones and the permafrost, and sequenced plant DNA directly from ancient soils. The picture that these data paint is of a dynamic and resilient ecosystem, where the dominant members of the community change along with the climate. During cold intervals, grasses are sparse, and horses and mammoths are more abundant than bison, presumably because they are better adapted than bison to surviving on poorer-quality forage. But when the climate warms, everything changes. Ground squirrels move in, grasslands expand, and bison rebound, outcompeting the other herbivores, possibly because bison have shorter generation times and can more quickly turn the reestablishing grasslands into more bison. The community is in constant flux; species come and go, become more or less abundant, and adapt and diversify in concert with each other and with the shifting climate.

Although our data are only from North America and go back only 40,000 years, I imagine that this scenario played out across Beringia and throughout the Pleistocene. During interglacial periods, habitats were filled with warm-adapted plants and animals. When the climate cooled, cold-adapted species took over while warm-adapted species became restricted to patches of refugial, warmer, habitat. In their refuges, the warm-adapted species eked out a perilous existence. Those individuals that survived became the foundation for new expansions once the climate warmed up again.

The seesaw between warm and cold periods occurred throughout the Pleistocene, instigating repeated cycles of expansion and decline in megafauna populations coincident with the waxing and waning of suitable habitat. Yet only the most recent climate transitions coincided with widespread extinctions of taxonomically diverse megafauna. Mammoths, which thrived in temperate, boreal, and tundra habitats on three continents for millions of years, suddenly found themselves with nothing to eat and nowhere to go after the Younger Dryas. Short-faced bears, which made it through at least two previous glacial cycles, were somehow unfit in the warmth of the Early Holocene. And horses, which survived more than a dozen major climatic transitions during the Pleistocene and thrive across the North

American West today, were unable after the last ice age to find suitable habitat anywhere in North America and went locally extinct. These and many of the other three dozen or so American species that became suddenly extinct after the last ice age survived multiple previous periods of similarly major climate change. The timing and coincidence of these extinctions demand an additional stressor. And what was different is that people were there, competing with other species for resources and refugial habitat and hunting them for their own sustenance.

The extinction of the North American megafauna fundamentally changed the North American landscape. On the North American plains, the disappearance of mammoths and horses paved the way for the reemergence and success of bison. Along the California coast, the disappearance of the megafauna instigated the return of closed shrubland and the decline of the California hazelnut, which was an important component of the local people's diet. To reopen the woodlands in the absence of megafauna, people adopted fire as a means to control the vegetation. In Beringia, the productive steppe tundra was replaced by the less productive tundra ecosystems of today, at least in part because the large animals that once recycled nutrients, dispersed seeds, and turned over the shallow soils are no longer there.

The veracity of Martin's blitzkrieg model is still debated. Arguments against it call into question the capacity of so few humans to kill such a diversity of large prey, and point to previous periods in Earth's history when dramatic ecological changes caused mass extinctions. To Martin, however, a climate change model was insufficient not only because it ignored the fact that similar climate changes had not caused mass extinctions during previous glacial cycles but also because it did not provide a common cause for the global nature of the extinctions. A compromise view has emerged in which the extinctions are seen to be caused by a combination of effects: climate changes disrupted habitats just as human populations were expanding, and as human populations grew so did their need for food and the intensity of their hunt. Martin also rejected this compromise view, again because it was not sufficiently global in its explanatory power. Yes, major climate changes coincided with the megafaunal extinctions and human arrival in the Americas, and also possibly in Europe and in northern Asia. However, there is as yet no evidence of significant climate change immediately preceding the Australian megafaunal extinctions.

CONTINUED ONSLAUGHT

The Polynesian ancestors of the Māori first settled on Aotearoa (also known as New Zealand) during the late thirteenth century, some 700 years ago. By 600 years ago, an entire order of birds—three taxonomic families that included nine different species—that had flourished on the islands for more than 5 million years was extinct.

Moas, the common name given to this diverse group of birds, were giants. A single moa egg was equivalent in size to around 90 chicken eggs. The females of the largest species—female moas were much bigger than male moas—weighed around 250 kilogram (500 lb). And moas had only one predator: the giant Haast's eagle, *Harpagornis moorei*, which could swoop out of the sky, pick up a moa with its giant talons, and carry it away.

Genetic data isolated from hundreds of moa bones, feathers, and eggshell pieces tell a story of moa populations that were large and thriving for at least the 4,000 years that preceded their extinction. There is no genetic evidence of population decline, illness, struggle to find food, or other hard times during the period leading up to their extinction. There is no paleoecological evidence of abrupt climate changes on the islands of Aotearoa coincident with or immediately preceding the moa extinctions. The climate on the islands had changed during the Pleistocene and Holocene, just as it had everywhere else, but moas weathered these changes with little genetic evidence of any impact. And then, in a geological instant, Aotearoa's moas disappeared.

What caused the sudden change in moa fortune? The answer should be pretty obvious by this point in the chapter. The earliest archaeological sites on Aotearoa are replete with moa bones and eggshells. This evidence of immediate and extensive use of moas by people makes moa extinction one of the easiest to tie directly to us. Just as Martin's model predicts, people arrived on the islands of Aotearoa and then ate all the mom moas, all the dad moas, and all the baby moas. The moas had not evolved with this type of predator and were unequipped to escape. The moas could not reproduce sufficiently quickly to sustain human appetites, and now they are gone.

The story of the Aotearoan moa extinctions is similar to other island extinction stories, and these stories differ from those of continental extinctions in a couple of important ways. First, megafaunal extinctions on islands tended to occur more quickly than did extinctions on the continents, with less time between the first arrival of humans and the death of the last of the megafauna. This probably comes down to the size of islands (species have less available habitat in which they can hide than they do on continents), the size of prey populations on islands (smaller populations are more vulnerable to becoming extinct than are larger populations), and the nature of the prey species themselves (many island species evolved in the absence of any predators of any sort). In addition, because island-dwelling humans could also take resources from the ocean, declines in land-based prey populations would have had little limiting influence on the size of the predator (human) population. The second way that island extinctions differ from those on the continents is that island extinctions occurred later in geological time. The reason for this is simple: it took people longer to get to the islands.

The first appearance of humans on an island was almost always bad news for endemic fauna. Nearly 10 percent of native bird species went extinct as humans found and colonized the Pacific Islands. As elsewhere, species with larger body sizes and slower rates of reproduction were the most prone to becoming extinct. Although birds were particularly vulnerable, island extinctions were not limited to a single taxonomic group. Dwarf hippos disappeared from the island of Cyprus 12,000 years ago, shortly after the first humans appeared. West Indian sloths survived on the islands of Cuba and Haiti longer than they did on the mainland, with the timing of their initial population declines on the islands closely matching the first archaeological evidence of people. The Jamaican monkey, which was the last surviving Caribbean primate, went extinct within 250 years of human arrival on Jamaica.

The extinction of island species was not always due to the direct effects of human predation. The arrival of people was often accompanied by land clearing and other types of environmental degradation, which reduced habitat for endemic species. Seafaring people also brought things with them, either because they liked those things—edible plants, domesticated animals like dogs, pigs, and chickens—or because they weren't aware that those things were coming along. Rats have been particularly devastating to island species. Rats were sometimes transported as food and sometimes hitched a ride unbeknownst to their captains. In fact, the dispersal of rats is so closely tied to the dispersal of people that a genetic analysis of Pacific rats has been used to reconstruct the timing and order of human settlement of the Pacific Islands.

One of the last islands to be colonized by people also happens to have been the home of one of the most famous now extinct species. The dodo, a flightless pigeon with the dubious honor of being the global icon for human-caused extinction, was endemic to Mauritius, a small island in the Indian Ocean about 1,200 kilometers (750 mi) from Madagascar. The earliest written record of dodo on Mauritius is from Portuguese sailors who discovered the island in 1507 after being blown off course by a cyclone. The sailors didn't stick around, and none of their writing mentions the dodos. In 1638, Dutch sailors and tradesmen established the first permanent settlement on Mauritius. Twenty-four years later, dodos were extinct. Stories of the dodo extinction depict it as among the cruelest and most brutish of human actions. Dodos were silly-looking birds that, rather than run away, would wander directly up to a person to check him out. Some books record people clubbing dodos to death for sport or amusement. People did not, however, eat dodos, which apparently didn't taste particularly good. Instead, dodos succumbed to an inability to reproduce. Female dodos laid a single egg in a nest on the ground during each breeding season. These eggs were gobbled up by the rats and pigs and other species that humans brought with them. No baby dodos were born, and eventually all the adult dodos died.

Are humans responsible for the deaths of the dodo, the moas, the Cyprus dwarf hippo, West Indian sloths, and the Jamaican monkey? The weight of evidence is against us. But there are a few counterexamples. A 4,200-year-old West Indian ground sloth tooth was recently found on the island of Cuba, proving that this species coexisted with humans for more than 3,000 years. While this does not mean that people are off the hook for their extinction, this proof of long coexistence demands a mechanism other than blitzkrieg.

Although island fauna seem to have suffered disproportionately compared to those on the continents, the small size and low density of potential prey on islands also led to some of the earliest human efforts to conserve animal resources. In Sri Lanka, humans have been hunting toque macaques, gray langurs, and purple-faced leaf monkeys for the last 45,000 years, and yet all three species are still around today. Patrick Roberts of the Max Planck Institute for the Science of Human History in Jena, Germany, studies the interactions between indigenous Sri Lankans and the native primates, and believes that the only reason these primates survived to today is because of deliberate actions on the part of the early Sri Lankans. In Fa-Hien Lena Cave, Roberts and colleagues identified thousands of bones from hunted animals, most of which represented adult individuals—despite the fact that adults in the prime of health would have been the most elusive in the population. Roberts believes that this preference for adults reveals a sophisticated hunting culture that developed considerable knowledge about the breeding cycles and territories of their prey. The Sri Lankan hunters, he argues, were sufficiently aware of how hunting might affect the primates to deliberately limit their take. If true, the early Sri Lankans may have been the first humans to practice sustainable hunting.

LONELY GEORGE

On New Year's Day 2019, a small and rather unremarkable-looking tree snail died in a captive breeding facility at the University of Hawai'i at Mānoa. The snail, named George, was born 14 years earlier and had lived in the facility for his entire life. After all his relatives died, researchers searched the island for a mate, or at least a friend, for George but never found another snail of his species. When George died on January 1, 2019, his species, *Achatinella apexfulva*, became the latest to move into the "extinct" category of the International Union for Conservation of Nature (IUCN) Red List.

While stories about the extinction of a Hawaiian tree snail are less likely to attract attention than stories about the impending extinction of something bigger, like a rhinoceros, the losses of less conspicuous species are no less damaging to their ecosystems, and no less our fault, than the losses of large herbivores. In fact, some scientists estimate that nearly 40 percent of species that have become extinct since the 1500s are land snails and slugs. In Hawai'i, snails play vital roles in their ecosystems. Some act as decomposers, occupying the niche filled by earthworms on the mainland. Others eat the algae that grows on leaves, potentially limiting the spread of disease. But, thanks to snail predators that have been introduced to the Hawaiian Islands, endemic Hawaiian snail species have been disappearing at an alarming rate. Introduced rats eat snails, and people like to collect and sometimes eat snails too. But the worst offender is another snail: the rosy wolfsnail.

Rosy wolfsnails were brought to Hawai'i in 1955 to counter the spread of giant African land snails, which were brought to Hawai'i accidentally in 1936 and have since been eating their way through crops and plaster walls across the islands. Rosy wolfsnails are voracious predators of other snails and were intended to gobble up giant African land snails and solve the problem. Unfortunately, rosy wolfsnails prefer the taste of Hawaiian snails, and today, while giant African land snails are doing just fine, those endemic snails that are not yet extinct are well on their way. The final blow to Hawai'i's endemic snails may come from the coincidence of our well-meant intervention and the changing Hawaiian climate. Some endemic Hawaiian snails can be found today in refugia at higher altitudes, where the climate has been too cold and too dry for rosy wolfsnails. These habitats are, however, becoming both warmer and wetter, and rosy wolfsnails are moving in.

Although George's story does not have a happy ending, the plight of the Hawaiian snails perfectly captures the present wrestling with the transition between our species' role as predator and that of protector. We know why some species are disappearing and want to stop this from happening, but we don't yet have a solution. We can establish captive breeding programs, but if there is no mate or if the species won't breed in captivity, then there is nothing else to do. We can design and implement strategies to remove invasive species, but those strategies might fail or have unintended consequences. And while we imagine newer ways to help endangered species, their habitats continue to deteriorate.

New technologies are on the horizon. If George had been sufficiently closely related to a different species of snail, perhaps they could have mated and produced offspring while George was still alive. This offspring, however, would have gotten only 50 percent of its ancestry from George's species, and it is unclear whether it would be considered by taxonomists or conservation biologists as the same species or as something else. This outside genetic input might also change the snail's behavior. If the hybrid didn't fill the ecological niche that George's species once filled, then the consequences of extinction would not have been averted.

It might be possible someday to clone George. In fact, a small piece of his foot was cut off just after he died and sent to the San Diego Frozen Zoo, where it is cryopreserved—just in case the technology to resurrect an extinct species comes to fruition. De-extinction, however, is not an imminent solution to the extinction crisis and is a particularly distant option in the specific case of George. De-extinction requires viable cells that can be transformed into viable embryos, a maternal host capable of providing an environment in which the embryo can develop, and a reliable strategy to rear and release a cloned animal in the absence of any individuals belonging to the same species. Every candidate species for de-extinction will face different technical, ethical, and ecological hurdles along the way, ranging from "Can we find a viable cell?" to "Is there an ideal egg or maternal host?" to "Is there a habitat into which we can release the organism without it immediately becoming re-extinct?" In George's case, it may be that the cells in storage at the frozen zoo are viable. However, biologists know too little about the reproductive and developmental cycles of George's species (and snails in general) to develop an effective cloning and husbandry strategy. This, of course, may change with time. But I would wager that snail-cloning technology development is low on the list of biotechnologies in need of development, and that private investment—the type of funding that powers these proposed solutions to the extinction crisis—is more likely to go toward resurrecting charismatic extinct megafauna than to bringing back a tiny snail. I am not hopeful for George's de-extinction, but his foot is frozen, just in case.

AN EVOLVING FORCE OF EVOLUTION

Of all the ways that humans have messed with the evolutionary trajectories of the species around us, causing them to become extinct is by far the most poignant. Perhaps this is why we go to such great lengths today to excuse ourselves from culpability, to find other explanations or partial explanations, and even to imagine technological solutions like de-extinction that might absolve us of our responsibility. Rather than pass the blame, however, it would be more fruitful to recognize that the extinctions that humans caused or helped to cause in the past were not deliberate actions. The first Australians and first Americans did not decide and then go out of their way to kill all the giant wombats and all the mammoths. Rather, the arrival of humans fundamentally changed the selective landscape of these habitats. Just as bipedal apes had a fitness advantage compared to their quadruped cousins when hunting in fields of tall grasses, animals whose behavior and physiology kept them out of our Pleistocene crosshairs had a fitness advantage over those that did not run away. Even later extinctions, such as those of the moa and dodo, were not deliberate. These and many other now extinct species were victims of dramatic changes to their habitats, some of which people caused directly, others indirectly, and some of which were perhaps not our fault at all.

It is clear, however, that the severity and destructiveness of human actions increased over time. This is in part because there were increasingly more of us. Instead of a few small subsistence-hunting family groups, people arrived on islands by the boatload, and when they disembarked from their ships, so did rats and cats and pigs, and they brought seeds of domesticated plants, and insect pests and giant African land snails. Over time, people also became increasingly powerful. Rudimentary tools were replaced by atlatls, and then shotguns, and then supercomputers. As our technology advanced, that technology accelerated the pace of further technological advance. We developed new ways to kill and to transform landscapes, and did so, and continue to do so, at a pace far faster than natural selection—the only mechanism by which other species can adapt—can keep up. The Sixth Extinction—the mass extinction that is undeniably underway today—is the consequence of human evolution.

The news, however, is not all bad. In addition to technological prowess, we have evolved a social conscience. We don't want to drive species to extinction. We want to protect habitats and preserve biodiversity. For some, the desire to conserve nature is entirely selfish: they enjoy the aesthetics of natural spaces and the variety of choices on the menu. Others are more altruistic, ascribing value to nature for its own sake. Regardless of motivation, the most recent 150 years has seen a gradual recognition of our role in causing extinctions and an equally gradual adoption of a new role in which we work actively to not cause any more extinctions. But as we embrace this role, it is clear that we cannot step back and allow species to evolve along their prehuman trajectories. We are too embedded, our technologies too advanced, our population too large to disentangle ourselves from the habitats that we have invaded over the last 200,000 or so years. Instead, the new human role is to consider how our actions, beginning in the Pleistocene, affected other species and to learn from the consequences of our past. We must consider what a world that is both biologically diverse and filled with people might look like. We must use our increasingly advanced technologies to shape the future into one in which people can thrive alongside other species.

CHAPTER 4

Lactase Persistence

O N THE COPY OF CHROMOSOME 2 THAT I INHERITED FROM my mother, around two-thirds of the way toward the centromere on the chromosome's long arm, I have a single mutation—a change from a guanine (G) nucleotide to an adenine (A) nucleotide—in intron 13 of a gene called *minichromosome maintenance complex component* 6, or MCM6. MCM6 is involved with making the protein complex that unwinds DNA during cell division, which sounds (and undoubtedly is) important. My mutation doesn't affect this process at all, though, as my mutation is in an intron, which is a string of DNA that is ignored when the gene is translated into a protein. Given that introns end up on what is essentially the genomic cutting-room floor, it might be surprising to learn that this particular mutation, when it occurred in one of my ancestors, changed the course of human evolution.

The G to A mutation in intron 13 of MCM6 has only existed in humans for a short time, and yet a lot of people have it. I have one copy, which I inherited from my mother, but many people with northern European ancestry have two copies of the mutant version of intron 13, meaning that they inherited it from both of their parents. In Europe, the prevalence of this mutation increases in frequency on a cline from south to north. More than 60 percent of people living in central and western Europe carry at least one copy of the mutant version of MCM6, and this proportion climbs to over 90 percent of people in the British Isles and Scandinavia. This pattern—a mutation that rises to high frequency in a big population over a short window of evolutionary time—is one that cannot emerge by chance. This mutation must have provided a fitness advantage to those who inherited it. In fact, when the fitness of people with the mutation is measured against the fitness of those without it, this mutation emerges as the most strongly advantageous mutation to have arisen within the last 30,000 years of human evolutionary history.

Given how important the G to A mutation in intron 13 of MCM6 appears to have been, one might guess that scientists would have some deep understanding of exactly what it does. The phenotype caused by this mutation is well described: Someone who has this mutation retains the ability to digest lactose, the type of sugar found in milk, into adulthood. Normal people (meaning those who don't have this mutation) lose the ability to break down and digest lactose near the age of weaning, just as all other mammals do. Non-mutant adults are "lactose intolerant"—drinking milk might make them bloated and gassy. Mutant adults, however, can drink gallons of milk without any unpleasant side effects. Well, maybe not gallons.

The molecular mechanism leading to "lactase persistence," which is the scientific name describing continued production of the lactase enzyme into adulthood, is less well understood. The G to A mutation in intron 13 of MCM6 occurs 14,000 nucleotides away from the lactase gene, which leads one to wonder how a mutation so far away can have any impact at all. The mutation appears to alter the sequence of intron 13 in such a way that it becomes a secondary binding site for the protein that turns on the lactase gene. With this secondary activation site, lactase continues to be made after the normal pathway turns off, and the person continues to be able to drink milk.

Exactly when lactase persistence first evolved in humans remains a mystery. The archaeological record shows that dairying spread quickly after it appeared but has not yet revealed whether these first dairying populations were already lactase persistent. It may be that the mutation emerged later, after dairying was already well established. In addition, genetic data from human populations from around the world have revealed several different mutations that cause lactase persistence and that arose independently. Strangely, the commonness of lactase persistence in a population does not correlate absolutely with cultural dairy use. The high frequency of lactase persistence in some present-day populations may be due simply to chance: a convergence of events that created the ideal environment for the trait to spread. Regardless of how or when lactase persistence mutations appeared in our genomes, however, it is clear that these mutations not only altered the evolutionary trajectory of our own species but also cemented the transformation of many other species from wild to less so—a transformation that continues today.

The story begins at the end of the last ice age.

HUNTERS BECOME HERDERS

By 14,000 years ago, the most recent ice age was over and humans had spread across much of the world. The warmer and wetter climate meant ever higher yields of edible plants and well-fed prey. With plenty to eat within easy reach, people began to trade life on the move for a more sedentary existence.

One particularly productive part of the world was the Fertile Crescent, a crescent-shaped (as the name suggests) region that extends from the coastal Levant up and then down the foothills of the Taurus and adjoining Zagros Mountains and to the Persian Gulf. Residents of early Fertile Crescent communities hunted wild animals and gathered fruits, seeds, leaves, and tubers that flourished around them. Securing the next meal was easy, and so they began to experiment. They discovered that selectively taking male rather than female prey animals not only provided more bang for the buck but also allowed prey populations to grow. They found a reliable and nutritious source in the increasing number of nut-bearing trees, grasses, and legumes that proliferated under these warmer and wetter conditions. They developed tools—grinding stones—to help them process the bounty of nuts that they gathered. It was a time of abundance and innovation.

Then, suddenly, fortunes changed. Around 12,900 years ago, the planet plunged back into ice age conditions—the Younger Dryas. People found themselves in a much less productive landscape and, having spent the good times transforming carbon and nitrogen from the ground into more humans, with more mouths to feed. Despite the downturn, conditions in some parts of the Fertile Crescent were still pretty good compared to elsewhere, and people concentrated there. The bad times lasted more than a thousand years.

When the climate rebounded and the Holocene began, the communities of the Fertile Crescent once again began to experiment. They capitalized on the natural resources from their land and transformed the innovations of their ancestors into new strategies that ensured survival. As they experimented, the plants and animals with which they interacted began to change. In this new era of human experimentation, the fittest individuals were those that humans chose to propagate. This was the dawn of the Neolithic.

The word *Neolithic* translates literally into "New Stone," and refers to the New Stone Age, or the period of human history during which plant and animal domestication began. The transition into the Neolithic started toward the end of the thousand-year Younger Dryas downturn and, by 10,000 years ago, people were planting, tending, defending, and harvesting crops of wheat, barley, lentils, peas, flax, and chickpeas, among others. These crops required year-round attention, which meant more people were needed to fill these roles, and more food was needed to feed these people. Wildland was transformed into farmland. Small communities became agricultural villages, and then villages became towns and then cities. More resources were taken from the ground to create and maintain the infrastructure for these settlements, and more land was transformed from wild to cultivated to feed the people who supported this infrastructure.

The shift from gathering edible plants to cultivating crops provided some security against starvation, but crop yields varied. With many more mouths to feed, lean years would have been awful. People needed a way to cache food so as to protect themselves against starvation in years of poor harvest. They built infrastructure for storing harvested grains, but this attracted rodents and other pests and the stored food didn't last forever. Fortunately, they had a solution for this, too. Animals, which people had also begun to manipulate by this time, were a reliable vessel for storing calories. During good years, excess grains could be fed to captive animals, whose populations would grow. These animals could then be harvested when crop yields were poor or whenever people fancied a bit of meat in their diet.

The transition from hunting to herding was a slow process of trial and discovery. People fiddled first with limiting animal movement, understanding that contained prey are easier to catch than are prey dispersed across wide distances. They learned to interpret animal behavior, experimenting with selecting based on temperament which individuals to eat and which to allow to reproduce. They saw that some species fared better than others around people. Some species were less likely to run away or to be nervous in captivity, for example, and species that naturally followed the lead of a dominant individual were more inclined to take similar cues from a dominant person.

As with plants, the earliest evidence that humans were manipulating wild animals is from the region around the Fertile Crescent.

Around the beginning of the Holocene, hunters just northwest of the Fertile Crescent started to use a new strategy to hunt the wild ancestors of domestic sheep. They began to almost exclusively take young reproductive rams—males between the ages of two and three. This hunting strategy would have had two benefits. First, taking males rather than females would ensure that the herd could continue to reproduce. Second, the targeted removal of local males may have attracted males from nearby herds. This new strategy essentially allowed the hunters to have their sheep and eat them too.

By 9,900 years ago, goat herders at Ganj Dareh in the highlands of present-day Iran had inherited a similar strategy to maximize returns from the hunt, taking reproductive males and older females and leaving reproductive females to replenish the herd. Then, some 500 years later, goats appear in the archaeological record of the nearby lowlands. This marks an important transition in human history. The highlands are a natural mountainous home for wild goats, which spend their time climbing up precarious ledges to escape predators and acquire food that other animals can't reach. The lowlands, however, were not particularly good goat habitat. Goats were in the lowlands because people had brought them there. These goats were on their way to being domesticates.

TRANSFORMATION

There are no rules that define the transition from wild to domestic. We define domesticated species as those whose evolutionary trajectories are within human control. Today, people choose for our domesticates which female mates with which male and which seeds to plant and which to discard. If we choose wisely—drawing from thousands of years of breeding expertise and decades of experiments in genomics—our choices will change how that species looks, acts, or tastes, and that species will be transformed according to our vision. Of course, our early Neolithic ancestors had no experience in the deliberate manipulation of other species. For them, the transformation of a species from wild to domestic was, at least initially, serendipitous.

Consider our first domesticate: the dog. Genetic evidence tells us that dogs were domesticated in Europe or Asia by at least 15,000 years ago and possibly much earlier. There was, however, no ice age hunter who thought it'd be fun to bring a wolf inside his hut to let it sleep on his bed and keep his feet warm. Instead, the transition from wolf to dog began by accident, when wolves denning near a human settlement recognized people as a potential source of food. They weren't eating people, of course. If they had, our ancestors would have killed them and the whole "man's best friend" thing never would have happened. Instead, wolves were cleaning up—scavenging stuff that people discarded and catching and eating other species that were also scavenging discarded stuff. The initial relationship between wolves and people was not mutualistic but commensal: wolves fared well in proximity to people, and people may or may not have noticed.

The stage was set, however, for this interaction to intensify. Wolves stuck by human settlements and the people living in those settlements started to benefit from their presence. Wolves denning nearby ate food waste, which kept away pests. If more dangerous predators approached, startled wolves would provide an early warning. And wolf puppies were probably pretty cute, biologically speaking. As the relationship matured, wolves and people gradually lost their fear of each other and a two-way mutualism developed. Humans fed and bred those wolves that were the least aggressive and the least likely to run away. Eventually, these wolves evolved into dogs, and people came to rely on them as workers and companions.

Dogs were the first but not the only species to follow the commensal route to domestication. Cats, which were among the earliest species to be domesticated, were also attracted to the trash that human settlements produced. Well, they were attracted to the mice and rats that were themselves attracted to the by-products (trash) of early farming. Unlike dogs, cats did not undergo extensive physical changes during their transition from wild to domestic, but they are tame compared to wildcats. Cats, like dogs, have evolved to be particularly tuned in to human social cues. Some cats clearly recognize and react to their names, for example, and others will follow nonverbal instructions such as to select between options based on where their owner points. If they can be bothered, that is.

As our companions, dogs and cats both hold elevated positions in human society—we tend not to eat them, for a start—but not all domesticates that followed the commensal pathway landed such cushioned roles. In China, wild jungle fowl that were attracted to discarded food scraps evolved into domestic chickens. Domestic chickens now account for 23 billion of the 30 billion land animals living in farms across the world, sometimes packed so closely together in cages that they can barely move. Wild turkeys were domesticated in present-day Mexico and the southwestern United States, and human preference for large turkey breasts has created some lineages that are too top-heavy to walk or reproduce on their own. And pigs, which are believed to have entered their relationship with us as scavengers of trash in Southwest and East Asia, are today farmed for meat, kept as pets, used for replacement human body parts, and derided by cultures across the world by association with stereotypes that are almost always negative.

Most species that we think of as livestock—cattle, sheep, goats, camels, buffaloes, and others—came under domestication via a second pathway: the prey pathway. The prey pathway is similar to the commensal pathway in that it begins by chance, although the first stage in the prey pathway is human management of wild animals. Animal management may have been triggered by a local scarcity of prey, perhaps due to climate change or overexploitation or both. Or perhaps animal management emerged as people fiddled with hunting strategies in ways that enhanced both numbers and predictability of prey. Either way, the strategies that were developed—taking post-reproductive individuals or harvesting males rather than females, for example—sustained and even boosted prey populations and helped to ensure a steady supply of food. Eventually, some managed prey species were further co-opted into human society. People began to control where and when they moved, what they ate, and how they reproduced.

As these animals traded life in the wild for life under our control, they experienced a different suite of evolutionary pressures. In captivity, horns were not only unnecessary for protection or mate competition but were also energetically costly to grow and carry around. More important, however, was temperament. An aggressive captive animal was a danger both to people and to other animals and could not be tolerated. An animal that spooked easily and ran away would also not contribute to the next generation. Over time, humans chose to breed the most docile and subdued animals, creating herds that were no longer fearful of, and even came to rely on, the herdsmen. Today, some scientists believe that many of the seemingly unrelated physical traits that are common among domesticated animals—piebald coat color, small teeth, floppy tails and ears, small brains, and nonseasonal estrous cycles, for example—stem from genetic changes in the developing brain that are also associated with selection against aggressive behavior.

Although the commensal and prey pathways begin unintentionally, there is a point along both pathways after which our choices become deliberate. During the first generations of life in captivity, horns may become smaller and individuals more docile because individuals with these traits are more fit than others in their captive environment. However, when a herder decides that he prefers animals of a particular size or coat color, or that he wants to create an animal that can pull a plough or produce lots of milk, a line is crossed. This transition from chance to intention is what sets domestication apart from other mutualisms, and what sets us apart from other animals.

Mutualisms that look a bit like domestication are common across the tree of life. Some of the most intriguing examples are in ants. Tropical leaf-cutter ants follow well-worn paths across the forest, each individual carrying a piece of leaf many times its size. The leaves are not intended for the ants to eat but instead to feed to the giant crops of mushrooms that the ants cultivate in their colonies. The ants in this mutualism are much like human farmers. They feed and prune their fungal gardens, keeping them healthy by removing parasitic fungi and other pests. They even learn (via chemical cues) whether a particular plant is toxic to the fungi and stop providing it. The fungus benefits by being provided safe habitat underground—habitat unavailable to fungi outside the mutualism. The mutualism is passed on to the next generation thanks to another striking leaf-cutter ant behavior: On a day that begins otherwise normally, thousands of winged ants suddenly take to the sky en masse. Once airborne, they mate several times and then lose their wings and fall to the ground like black snow with pinchers. This hellscape (which I happened to experience as an undergraduate working in the rain forest in Panama) is not a modern-day plague but instead the nuptial flight of the leaf-cutter ant. And the fungus? Throughout all this flying and frolicking and falling, each hopeful queen of the next generation manages somehow to hold tight to a bit of her old colony's fungus. If she survives, she will use this to sow her own garden of mushrooms, passing on the mutualism.

Some ants have evolved mutualisms that look more like herding than farming. European yellow meadow ants, *Lasius flavus*, tend aphids, protecting the tiny insects as the aphids feed on plants. In return, the ants eat the aphids' nutrient-rich poop (some call this "honeydew"), which the ants coax the aphids to release by stroking them with their antennae (some refer to this as "milking"). The rare African ant *Melissotarsus emeryi* protects colonies of scale insects, squeezing and licking them, presumably eating the wax that coats their bodies. *Melissotarsus* ants have also been observed plucking scale insects from plants and carrying them away, raising suspicion that the scale insects are also sometimes the dish of the day.

Other animal mutualisms mimic pet-and-owner relationships. The giant venomous tarantula *Xenesthis immanis* allows a tiny frog— the dotted humming frog *Chiasmocleis ventrimaculata*—to hang out in its burrow. While the tarantula could eat the frog if it wanted to, it doesn't. Instead, the frog just hangs out in the burrow with the spider. After the spider finishes a meal, the frog steps in to gobble up whatever's left behind. The frog gets a safe place to live, and the tarantula gets a burrow that is free from any leftover food scraps that might attract spider egg–eating pests.

On the surface, these mutualisms seem remarkably similar to domestication: farming, herding, ranching, even keeping pets. There is, however, a key difference. Domestication involves purpose. These ants, spiders, and frogs arrived in their relationships by chance. Over thousands of generations, each lineage adapted to coexistence with the other, eventually coming to rely on the mutualism for survival. At no point during the evolution of their relationship did either species notice that they had entered into a mutualism. Neither species stopped to consider alternate future scenarios in which the mutualism might be improved by enhancing a particularly useful trait. Neither species intentionally modified the other.

People, however, both design and tinker. And we do it quickly. In a human lifetime, someone trying to corral a wild aurochs can try out many different strategies. When she discovers that the strategy she's using isn't working, she can change her mind and move immediately to a different strategy, which she can refine based on what she's learned about the behavior and natural history of the aurochs. When she discovers a strategy that works, she doesn't have to wait for evolution to pass that innovation down to her child and eventually her grandchild. Instead, she can tell her child what she's learned. She can also tell her parents, and her neighbors, and her friends, and they can start to tinker where she left off.

As our ancestors began more than 10,000 years ago to manipulate their environments and the species within them, they were not intending to domesticate anything. They did, however, have a goal. They experimented with hunting strategies that would sustain the herd, with propagation strategies that would produce high-yielding crops, because they foresaw a future in which resources would be more predictable and secured with less effort. With this future in mind, they became increasingly intentional in how they changed and manipulated species. They built barriers that sequestered their prey and created irrigation networks that watered their crops. And, because they could think on their feet, change their minds and strategies in an instant, and tell their friends and family what they discovered, they had the upper hand in the developing power relationships with these species.

A THIRD PATHWAY TO DOMESTICATION

Domestication did not change the rules of evolution. As with wild species, the fitness of a domesticate is measured as the number of viable offspring produced. As with wild species, successfully rearing offspring in captivity requires access to resources and to a mate. As with wild species, the evolutionary pressures of the environment in which a domesticate lives will determine what traits maximize that access. For a domesticate, however, we are those evolutionary pressures.

As soon as our ancestors realized that plants and animals could be manipulated, they wanted more: a wider range of plants to eat or use for clothes, tools, or building materials; a broader diversity of animals for food, work, and transportation. Armed with new understanding of plant cultivation and animal husbandry, people discovered more efficient ways to exert human preferences on domesticates, to refine species already under human management. Domestication came to be seen as an opportunity.

Species tamed via the commensal and prey pathways tended to have traits that made them amenable to human control. They lived in large social groups with dominance hierarchies that created roles into which they could insert themselves. They were sexually promiscuous and flexible in what they would eat, which meant a person could make these decisions for them. And they didn't scare easily in the presence of people. Some species, however, lacked these predispositions. To domesticate these species required a third pathway: the directed pathway.

Among the first species to be tamed via the directed pathway was the horse. The earliest evidence that people were managing horses comes from the Botai culture of Central Asia. There, archaeologists recovered 5,500-year-old horse bones with evidence of bit damage to the teeth—suggesting the horses were bridled or harnessed—as well as milk protein residues from clay pots—suggesting that people were drinking or processing mare's milk. As there's little chance that people would be milking a wild horse, this is pretty strong evidence that the Botai horses were domesticated.

For more than a decade, Ludovic Orlando, an ancient-DNA scientist based in Lyon, France, who specializes in horse domestication, has been leading an international collaboration (of which I am a part) to try to learn when, where, and how horses were domesticated. When our research team isolated ancient DNA from the Botai horses and compared it to DNA from living horses, we expected to find that Botai horses were the direct ancestors of today's domestic horses. To our surprise, however, they were not. Instead, the Botai horses contributed some genetic ancestry to today's Przewalski's horse, which is widely considered the only remaining wild horse. But Botai horse DNA has otherwise disappeared from the horse gene pool. The archaeological record is clear that the Botai people domesticated horses, but the lineage that they domesticated did not survive.

In 2019, hoping to identify the source population of living domestic horses, our team compared DNA from nearly 300 ancient horses from across the Northern Hemisphere to DNA from living domestic horses. We found evidence that people domesticated at least two other groups of horses, in addition to the Botai horses, both around 4,000 years ago. One of these domestication events took place in southeastern Europe and another in northern Asia. Like the Botai horses, however, neither of these domesticated horse lineages survived. In fact, we still don't know where or when people domesticated the lineage of horses that we know today.

Why did people of so many different cultures domesticate horses? It is widely believed that the cultures who first tamed horses did so to help with hunting other horses. Once horses were under human control, however, people discovered that they were useful for more than their meat and hides. Like cattle, horses could be milked, and horses could survive on much poorer-quality forage than that required by cattle. More importantly, however, horses could be ridden.

Riding horses provided many advantages. Corralling wild and domesticated animals was more efficient on horseback than on foot. The height advantage gained by riding a horse made it easy to dominate other people. And horses were tough and sure-footed and could traverse long distances over difficult terrain.

Horse riding transformed the cultures that first adopted it on the Asian steppe and then proceeded to transform human societies everywhere. In Europe, for example, nomadic people of the Yamnaya culture arrived on horseback some 5,000 years ago, their (newly invented) carts filled with the weapons and tools of their distinct culture: wheels, copper hammers, and a language that some linguists believe is the root of all Indo-European languages spoken today. As they entered Europe, the Yamnaya encountered settled farmers who had expanded northward out of the Fertile Crescent 4,000 years earlier. Horses, in this way, brought together the cultures that instigated the transition from the European Neolithic to the Bronze Age.

Today, people continue to take species from the wild and try to turn them into something less so. While some of us are satisfied with keeping dogs and cats, others want more unusual pets. During the Middle Ages, people of a certain status, bored of using run-of-the-mill cats to control their rodents, turned instead to genets, small African carnivores that look like a mix among a lemur, a cheetah, a ferret, and a kitten. Genets are not particularly cuddly but they are certainly exotic, and their popularity is resurging in the United States and Asia. Capybaras, large rodents native to Central and South America, are also increasingly popular as pets. While breeding in captivity has created capybara lineages that are more amenable to life with people, they require being kept in pairs as well as access to swimming pools, mud baths, and plenty of sunshine, which is a niche that most urban and suburban households cannot provide.

The directed pathway is also being used to create new types of food. Aquaculture is perhaps the most rapidly growing domestication industry globally, with nearly 500 marine and freshwater species brought under domestication since the turn of the twentieth century. The exotic meat industry is working to domesticate (or at least bring into captivity) wild animals ranging from alligators to ostriches to reindeer, with varying success. One example is American bison, which are becoming common on ranches in the United States and elsewhere. Ranched bison are a domestication success story; they are, for example, less aggressive and less likely to panic in the presence of humans than their wild cousins. Much of this domesticated disposition may, however, have come from crossbreeding with cattle. Crossbreeding is accepted, even encouraged, by the industry, and ranched bison can have as much as 37.5 percent cattle ancestry while still qualifying for sale as "buffalo." To help ranchers balance crossbreeding for manageability and maintaining sufficient bison ancestry, genomic tests have been developed to calculate the proportion of cattle DNA in a particular sire or dam.

Directed domestication is also important in the world of edible plants. Today, dozens to hundreds of plants are being transformed from edible to cultivation-worthy (reliable) and nutritious. The

American potato bean, for example, is a legume (which means it fixes nitrogen in the soil in addition to being edible) that makes lumpy underground tubers. The American potato bean was consumed by indigenous people across North America but never domesticated. It is, however, an excellent candidate. It is nutritious, simple to harvest, and grows well in (and improves) poor-quality soils. As a domesticated crop, the American potato bean could be planted in deteriorating habitats or in parts of the world where weather patterns are predicted to become more variable.

Plants that are already domesticated but have the potential to become better providers of nutrition are also targets of directed selection. New sunflowers are being produced that have larger and oilier seeds, for example, as are new versions of corn that can thrive in drought-prone habitats or resist disease. These efforts to manipulate both wild and domesticated plants to be more efficient and more reliable sources of nutrition may be our best chance to keep the world's population fed as the human population grows and the planet's climate patterns change.

CATTLE POWER

Aurochs, *Bos taurus primigenius*, evolved in present-day India more than 2.5 million years ago, as the planet cooled into the Pleistocene ice ages and grasslands began to replace once widespread forests. This shift in the dominant vegetation created new herbivore niches, in particular for animals that could take advantage of harsh but nutrient-rich grasses. Aurochs were large animals with short generation times, which made them among the most successful species in the expanding grasslands. By the time people were planting wheat in the Fertile Crescent, aurochs had spread across much of Europe, Asia, and North Africa, where they were hunted and eaten by people.

Around 10,500 years ago at a site called Dja'de in the Middle Euphrates Valley of the Fertile Crescent (today's northern Syria), the archaeological record reveals a sudden decrease in the size of harvested aurochs bones. The archaeologists working at the site imagined several explanations for this change. A shift to a hunting strategy in which people took mostly females could cause this size reduction, as female aurochs are much smaller than males. The harvest, however, included equal numbers of males and females, ruling out this possibility. A shift to taking only subadults could also cause this pattern, but the harvest included aurochs of all ages. Deteriorating habitat might lead to smaller adults, but the size reduction was more severe in male than in female aurochs, suggesting a cause that impacts one sex more than the other. Also, other species in the harvest did not appear to be affected, as one might expect if there were a common, external, cause. The reduction in the size of harvested aurochs, the archaeologists concluded, was the result of domestication. The hunters-turned-herders of Dja'de may have been allowing only the least aggressive and most easily manipulated males to reproduce. Because aggressiveness and size tend to go together in cattle, this selective pressure reduced the size difference between the two sexes and the size of the species overall. Dja'de, they concluded, contained the first evidence of aurochs management.

Only a few hundred years after Dja'de, similar changes are seen at another site, Çayönü, around 250 kilometers (150 mi) from Dja'de in the High Tigris Valley of present-day southeastern Turkey. At Çayönü, archaeologists observed changes both in the size of the harvested bones and in the stable isotopes of carbon and nitrogen in these bones, which revealed that, in addition to males becoming smaller, the aurochs' diet had changed. They had begun to eat plants that grew in open habitats, like farms, rather than in their native woodlands. Archaeologists did not see similar isotopic changes in bones of harvested red deer, another prey species with similar habitat and diet preferences to those of wild aurochs. This suggested that the change in diet was not caused by trends in the local climate but was instead specific to aurochs. The aurochs at Çayönü, the archaologists concluded, had adopted a lifestyle that relied on or at least tolerated humans. These aurochs were early cattle—specifically taurine cattle, Bos taurus taurus.

Although new archaeological data will certainly refine the story of taurine cattle domestication, Dja'de and Çayönü are the likely epicenters. Both sites are situated on relatively flat land compared to the surrounding mountainous regions, which would have been decent habitat for managing aurochs. They are the only sites in the region where people had transitioned to a sedentary lifestyle by 11,000 years ago, which may have been prerequisite for aurochs management. The sites were also close geographically, which would have facilitated the exchange of ideas, and perhaps even animals, between settlements.

Almost immediately after domestic cattle appeared in the ar-chaeozoological record, they began to spread. Cattle domestication came later than that of goats and sheep and (according to the ge-netic evidence) involved relatively fewer individuals, which perhaps speaks to how much more difficult they were to tame. Aurochs were larger than sheep and goats, more aggressive, and more difficult to handle and corral. Once tamed, however, cattle transformed human societies. Cattle could not only feed and clothe people but could be put to work. With cattle as our first work "horses"—pulling, towing, and carrying heavy objects—farms became more efficient. Inaccessible or difficult land could be cultivated. More nutrients could be taken from the land and turned into more people. With cattle to do the hard work of turning land into farms, people expanded out of their villages, taking with them the cultures, tools, and domesticated plants and animals that defined the Neolithic.

By 9,000 years ago, taurine cattle had spread into Europe, fol-lowing either the Mediterranean coastline or the Danube River. By

8,000 years ago they had spread into Africa. They dispersed across Africa's northern coast, then across the Strait of Gibraltar and into Europe again via the Iberian Peninsula. By 4,000 years ago, taurine cattle had reached central and northern China. Zebu, a different cattle lineage that was domesticated in the Indus Valley in present-day Pakistan around 9,000 years ago, spread into East Africa by 4,000 years ago and into Southeast Asia by 3,000 years ago. As both cattle lineages spread, they encountered and mated with wild aurochs, with previously established populations of domestic cattle, and with other related species. Some of this mixing occurred by chance but it was often deliberate, as Neolithic herders restocked their cattle herds by taking in other animals. The exchange of genes led to the emergence of locally adapted lineages. Tibetan cattle, for example, survive at high altitude thanks to mutations inherited not from other domesticated cattle but from native yaks.

As cattle adapted to their new environments, they also adapted to life with people. In 2015, a team of Irish ancient DNA researchers led by David MacHugh and Dan Bradley sequenced the complete genome of an aurochs that lived in present-day Britain before domestic cattle arrived there. They compared this aurochs' genome to the genomes of living domestic cattle, looking for genetic changes that evolved in cattle after they were domesticated. They found mutations in genes related to brain development, immunity, and fatty acid metabolism, which probably benefited cattle as they adapted to living in social groups and consuming different foods. They found mutations in genes involved with growth and muscle mass, which may reflect adaptations associated with meat production. And they found that European cattle share a mutation in a gene called *diacylglycerol O-acyltransferase 1*, or *DGAT1*, that has been associated with the production of particularly fatty milk. This mutation is evidence of early selection by humans for what today remains a vital part of our global food system: dairying.

LIQUID PROTEIN

The first archaeological evidence that people were dairying dates to around 8,500 years ago—2,000 years after cattle domestication. In Anatolia (present-day eastern Turkey), which is pretty far from the original center of cattle domestication, archaeologists recovered milk fat residues from ceramic pots, indicating that people were processing milk by heating it up. Similar analyses of milk fat proteins in ceramics record the spread of dairying into Europe, which appears to have happened simultaneously with the spread of domestic cattle.

It's not surprising that people began dairying soon after cattle domestication. Milk is the primary source of sugar, fat, vitamins, and protein for newborn mammals, and as such is evolved expressly to be nutritious. It would not have taken much imagination for a cattle herder to deduce that a cow's milk would be just as good for him and his family as it was for her calf. The only challenge would have been digesting it—without the lactase persistence mutation, that is.

Because lactase persistence allows people to take advantage of calories from lactose, it also makes sense that the spread of the lactase persistence mutation and the spread of dairying would be tightly linked. If the mutation arose near the start of dairying or was already present in a population that acquired dairying technology, the mutation would have given those who had it an advantage over those who did not. Those with the mutation would, with access to additional resources from milk, more efficiently convert animal protein into more people, and the mutation would increase in frequency. Curiously, though, ancient DNA has not found the lactase persistence mutation in the genomes of early dairy farmers, and the mutation is at its lowest European frequency today in the precise part of the world where dairying began. The first dairy farmers were not, it seems, drinking milk. Instead, they were processing milk by cooking or fermenting it, making cheeses and sour yogurts to remove the offending indigestible sugars.

If people can consume dairy products without the lactase persistence mutation, there must be some other explanation as to why the mutation is so prevalent today. And lactase persistence is remarkably prevalent. Nearly a third of us have lactase persistence, and at least five different mutations have evolved—all on the same stretch of intron 13 of the MCM6 gene—that make people lactase persistent. In each case, these mutations have gone to high frequency in the populations in which they evolved, indicating that they provide an enormous evolutionary advantage. Is being able to drink milk (in addition to eating cheese and yogurt) sufficient to explain why these mutations have been so important?

The most straightforward hypothesis is that, yes, the benefit of lactase persistence is tied to lactose, the sugar that represents about 30 percent of the calories in milk. Only those who can digest lactose have access to these calories, which may have been crucial calories during famines, droughts, and disease. Milk may also have provided an important source of clean water, which also may have been limited during periods of hardship.

Another hypothesis is that milk drinking provided access to calcium and vitamin D in addition to lactose, the complement of which aids calcium absorption. This might benefit particular populations with limited access to sunlight, as ultraviolet radiation from sun exposure is necessary to stimulate the body's production of vitamin D. However, while this might explain the high frequency of lactase persistence in places like northern Europe, it cannot explain why populations in relatively sunny climates, such as parts of Africa and the Middle East, also have high frequencies of lactase persistence. Neither this hypothesis nor the more straightforward hypothesis linked to lactase can explain why lactase persistence is at such low frequency in parts of Central Asia and Mongolia where herding, pastoralism, and dairying have been practiced for millennia. For now, the jury is still out as to why lactase persistence has reached such high frequencies in so many different parts of the world, and why it remains at low frequencies in some regions where dairying is economically and culturally important.

Ancient DNA has shed some light on when and where the lactase persistence mutation arose and spread in Europe. None of the remains from pre-Neolithic archaeological sites—economies that relied on hunting and gathering—have the lactase persistence mutation. None of the ancient Europeans from early farming populations in southern and central Europe (people believed to be descended from farmers spreading into Europe from Anatolia) had the lactase persistence mutation. Instead, the oldest evidence of the lactase persistence mutation in Europe is from a 4,350-year-old individual from central Europe. Around that same time, the mutation is found in a single individual from what is now Sweden and at two sites in northern Spain. While these data are sparse, the timing is coincident with another major cultural upheaval in Europe: the arrival of Asian pastoralists of the Yamnaya culture. Perhaps the Yamnaya brought with them not only horses, wheels, and a new language, but an improved ability to digest milk.

The mystery of lactase persistence in humans highlights the complicated interaction among genes, environment, and culture. The initial increase in frequency of a lactase persistence mutation, regardless of in whom it first arose, may have happened by chance. When the Yamnaya arrived in Europe, for example, they brought disease—specifically plague—that devastated native European populations. When populations are small, genes can drift quickly to higher frequency regardless of what benefit they might provide. If the lactase persistence mutation was already present when plague appeared and populations crashed, the mutation's initial increase may have happened surreptitiously. When populations recovered, dairying was already widespread and the benefit to those with the mutation would have been immediate. By domesticating cattle and developing dairying technologies, our ancestors created an environment that changed the course of our own evolution.

We continue to live and evolve in this human-constructed niche. In 2018, our global community produced 830 million metric tons (more than 21 billion US gallons) of milk, 82 percent of which was from cattle. The rest comes from a long list of other species that people domesticated within the last 10,000 years. Sheep and goats, which together make up around 3 percent of global milk production, were first farmed for their milk in Europe around the same time as cattle dairying began. Buffaloes were domesticated in the Indus Valley 4,500 years ago and are today the second largest producer of milk next to cattle, producing around 14 percent of the global supply. Camels, which were domesticated in Central Asia 5,000 years ago, produce around 0.3 percent of the world's milk supply. People also consume milk from horses, which were first milked by people of the Botai culture 5,500 years ago; yaks, which were domesticated in Tibet 4,500 years ago; donkeys, which were domesticated in Arabia or East Africa 6,000 years ago; and reindeer, which are still in the process of being domesticated. But those are just the most common dairy products. Dairy products from more exotic species—moose, elk, red deer, alpacas, llamas—can be purchased and consumed today, and rumor has it that *Top Chef*'s Edward Lee is working out how to make pig milk ricotta, should one want to try such a thing.

INTENSIFICATION

As the human relationship with cattle changed, so did the evolutionary pressures that our ancestors exerted on them. At first, people simply wanted cattle that were easier to control and to keep alive. These preferences led to a decline in cattle size (and presumably aggression) until the Middle Ages, when the trend reversed. Some historians have suggested that this reversal reflects a change in priorities during a time of prosperity. Free from worries about finding their next meal, cattle breeders of the time began to experiment with improving traits other than survival. By the seventeenth century, breeders had optimized cattle for the peculiarities of regional habitats and tastes, engineering breeds to be hardy in cold climates, sure-footed in mountainous terrain, and for aesthetic properties like coat color and horn shape.

Things were not, however, all sunshine and fields of clover. As cattle spread throughout Europe, they consumed the continent's resources. Forests and grasslands were transformed into cattle range, and species that relied on these wild habitats disappeared. Aurochs—cattle's wild ancestors—were among the species that cattle displaced; the last aurochs died in 1627 on a royal hunting reserve in present-day Poland. As cattle populations grew, the quality of cattle forage and habitats declined. Herds became densely populated and poorer in health. Rinderpest epidemics surged, causing chaos and famine across human society.

Searching desperately for a solution, cattle breeders turned once again to their engineering skills. They saw that some cattle were more resilient to disease than others and began work to blend this resilience with regionally advantageous and aesthetic traits. The stakes were high, and so they began keeping official breeding records, systematically acknowledging the blending of herds and results. This meticulous approach to cattle breeding eventually led to the diversity of cattle shapes, sizes, and varieties that we know today.

Cattle tinkering was not limited to European breeders. Cattle arrived in the Americas in 1493 as part of Christopher Columbus's second journey to the New World and, over the next hundred years or so, were brought from Spain, Portugal, and North Africa to the Caribbean and to Central and South America. The global cattle diaspora continued through the seventeenth century, as Anglo-Saxons brought European cattle breeds to North America and Australia. Zebu were introduced into Brazil from India during the eighteenth century. These new-to-cattle habitats presented new selective pressures and roles for cattle to fill. New World breeders refined their herds by combining traits among European breeds and adding traits from local animals like American bison, creating hardy and productive cattle that were better adapted than their European cousins to regional climates.

By the turn of the twentieth century, ranchers in Europe and the New World were struggling to keep up with demand, which had intensified with the invention of refrigerated railcars that could transport meat over long distances. As demand soared, so did efforts to maximize the profit potential of each animal. Breeders refined herds using data recorded in studbooks, sometimes importing animals from overseas to add to their herd's gene pool. Progress was slowed, however, by the limited number of offspring that any particular bull or cow could have. By the middle of the century, however, two new technologies—artificial insemination and embryo transfer—were on the horizon that would solve this problem and change forever the landscape of livestock breeding.

In artificial insemination, sperm are collected from a male and deposited into the reproductive tract of a female in estrus, allowing a breeder to control precisely which two individuals breed. More crucially, collected sperm can be frozen, shipped across long distances, and stored for decades while remaining viable. This increases the genetic impact of an individual bull far beyond his own reproductive years.

Embryo transfer similarly extends the genetic impact of an individual, in this case the cow. A typical female calf is born with more than 100,000 eggs, of which only a few will ever be fertilized. The goal of embryo transfer is to use more of these than the 10 or so that might be fertilized in her reproductive lifetime. In embryo transfer, cows are treated with hormones to induce more than one egg to be produced at a time. She can then either be mated (or artificially inseminated) to create more than one embryo, or her eggs can be collected and fertilized in vitro and then transferred to a surrogate host. If she is mated, her embryos are flushed out of her uterus and collected and, after making sure that they are healthy, transferred to surrogates to continue development.

Artificial insemination and embryo transfer expedited breed refinement. During the 1960s, breeders in Belgium used artificial insemination to refine an already strongly muscled cattle breed known as the Belgian Blue. Genomic analyses have since revealed that the extreme musculature of Belgian Blues is caused by a mutation that blocks production of myostatin, a protein that ceases muscle development when the animal reaches adult size. Animals with the mutation continue to grow even after reaching this developmental milestone, translating into enormous quantities of lean beef at slaughter. Although the genetic mechanism was unknown in the 1960s, breeders suspected that if a particularly muscular bull impregnated a particularly muscular cow, then she would give birth to even more muscular calves. Artificial insemination allowed them better control over their experiments by maintaining a constant male genetic input, expediting the pace of breeding selection.

Artificial insemination and embryo transfer have enormous agricultural potential. With embryo transfer, a cow can produce as many as ten calves per year compared to only one. This increased reproductive output can be lifesaving in parts of the world where frequent bouts of drought or disease lead to periods of food insecurity. These two technologies also make it possible to move desired traits quickly and efficiently between populations and breeds. In 1983, breeders from what is now known as the West African Livestock Innovation Centre transferred embryos from West African N'dama cows into Kenya Boran cow surrogates. Their goal was to transfer the innate resistance of N'dama cattle to trypanosomal disease—also known as African sleeping sickness—to impacted herds of Boran cattle. When the N'dama calves reached reproductive age, they bred with Boran cattle, passing their resistance genes to the next generation.

While proliferating a beneficial trait through a population can be a clear win, doing so via artificial insemination or embryo transfer can have unintended consequences. As an extreme example, if a particularly prized sire or dam becomes the only contributor to the next generation, this is equivalent to intensive inbreeding: all the resulting offspring are half siblings. Breeding selection has caused considerable losses of genetic diversity in most of our domesticates. In horses, for example, intensification of breeding selection within the last 200 years created the faster, more agile horses of today, but at the expense of a nearly 15 percent reduction in the species' overall genetic diversity. With fewer individuals in the breeding pool, what would otherwise be rare genetic defects increase in frequency. Belgian Blue cows, for example, have unusually narrow birth canals, which means that 90 percent of their (often extremely high-birth-weight) calves are born by Cesarean section. Belgian Blues are also prone to breathing problems and enlarged tongues, and their low fat production makes them poorly suited to living in cool climates. These animals would not survive outside human-controlled environments, and yet they are well adapted to the environment into which they are born.

Maladaptive traits that are side effects of breeding selection are common among our domesticates. Indeed some have become characteristics of particular breeds. A third of purebred bulldogs have severe breathing problems, for example, and German shepherd dogs are unusually susceptible to hip dysplasia. These genetic problems could be solved by introducing diversity by outbreeding, a process that is sometimes called "genetic rescue." Genetic rescue can, however, remove other breed-specific traits, which makes this solution sit uneasily with the dog-breeding community. In fact, this community went as far in the middle of the twentieth century as declaring genetic rescue "unnatural," although I note their acceptance as "natural" of the centuries of inbreeding that led to today's dog breeds. Fortunately, a controversy involving Dalmatians has brought many in the dog-breeding community closer to accepting genetic rescue.

By the 1970s, all purebred Dalmatians had a condition called hyperuricosuria, which caused urinary stones to form in the kidneys and bladder, often leading to kidney failure. In 1973, Robert Schaible, a medical geneticist and Dalmatian breeder, decided to try to cure his dogs of hyperuricosuria using genetic rescue. He bred a Dalmatian with a pointer, and then, to reestablish the proper Dalmatian breed traits, bred the resulting Dalmatian/pointer hybrids with purebred Dalmatians for five generations. Seven years later he had a line of dogs of 97 percent Dalmatian ancestry (31 of 32 fifth-generation ancestors were Dalmatians) that were hyperuricosuria-free. He petitioned the American Kennel Club to register his dogs as Dalmatians and, after several months of deliberation, his petition was approved. Then-president of the AKC William Stifel wrote, "If there is a logical, scientific way to correct genetic health problems associated with certain breed traits and still preserve the integrity of the breed standard, it is incumbent upon the American Kennel Club to lead the way."

Stifel's attitude was not shared by everyone in the Dalmatian-breeding community. Some community members pointed out that the problem had not actually been solved. Dogs could, after all, be carriers of hyperuricosuria without showing signs of disease and pass the trait to their offspring. In response, the AKC made breeders wait to register their dogs until their noncarrier status was proven, which was accomplished when the dogs produced entirely disease-free litters.

A simpler solution came in 2008, when Danika Bannasch from the University of California, Davis, identified the disease-causing mutation as a gene called SLC2A9. With this discovery, a simple genetic test could determine a puppy's carrier status at birth. In 2011, the AKC began registering Dalmatians that passed this genetic test, paving the way for genetic solutions for other diseases caused by breeding selection.

Genetic tests like that for hyperuricosuria have begun to change how breeding choices are made. Global DNA sequencing collaborations have created reference databases for bears, dogs, cattle, apples, maize, tomatoes, and dozens of other domesticated lineages. With these data, scientists and breeders are discovering which genes map to which traits and using this information to match breeds to environments, to move traits among populations and breeds, and to avoid the propagation of maladaptive genes. The consequences are significant: a 2019 report by Thomas Lewis and Cathryn Mellersh, both associated with The Kennel Club in the United Kingdom, found that several genetic diseases, including exercise-induced collapse in Labrador retrievers, early-onset cataracts in Staffordshire bull terriers, and progressive blindness in cocker spaniels, have been nearly eliminated among Kennel Club–registered dogs since genomic tests have become available.

TODAY'S DOMESTICATES ARE FEATS OF HUMAN INGENUITY AND engineering, and they are everywhere. According to the World Cattle Inventory (which is a real thing published by the website Beef2Live.com, whose motto is "Eat Beef, Live Better!"), our planet is home to nearly 1 billion cattle. This means there is one cow or bull for about every seven people and tons of cattle diversity. *Cattle Breeds: An Encyclopedia* describes more than 1,000 distinct breeds—three times the number of breeds of domestic dogs listed by the World Canine Organization. Most of today's cattle breeds emerged during the 250 years since the Industrial Revolution, as cattle were bred either for specific traits associated with meat or milk production or to live in unusual climates or locations. The global impact of all these animals has not been small. Cattle eat a lot. Each cow or bull can consume more than 18 kilometers (40 lb) of food every day. Cattle take up a lot of space. Land clearing to support livestock grazing has been a major source of anthropogenic landscape change over the last century. And cattle pollute. Methane gas released as their farts and burps (mostly burps) is responsible for nearly one-seventh of present-day global greenhouse gas emissions.

Yet we want more. We want better cattle with higher-quality meat and milk. We want different and more exotic pets. We want new varieties of tastier food. We want helpers and workers and guiders and sniffer-outers. At the same time, though, we are aware in ways that our ancestors were not of the consequences of getting what we want. We recognize that extinction rates of non-domesticated species are many times higher than the background rate in the fossil record, so we want livestock that produce more protein while requiring fewer resources and taking up less space. We know that industrialized agriculture causes air and water pollution, so we want our livestock to pollute less and our crops to be naturally immune to diseases and other pests. We see that some animals whose lives we control are suffering from genetic maladies made common by inbreeding, so we want to purge harmful mutations without sacrificing the traits that we've engineered. And we want to continue to refine our domesticates to improve our own lives, for example by engineering hypoallergenic cats or cows that produce milk drinkable even by those of us without a lactase persistence mutation.

These twenty-first-century goals are achievable with twenty-first-century technologies. The breeding strategies that our ancestors developed during the Neolithic and then refined over millennia are limited by the constraints of evolution: the randomness of genetic recombination and fertilization and the slow pace of generations. Today's DNA sequencing technologies can pinpoint specific genes that achieve a desired phenotype, but breeding is too imprecise to take advantage of this information. A new family of technologies loosely referred to as genetic engineering, and a new field of research—synthetic biology—that takes advantage of genetic engineering technologies, can improve our experimental precision. Using genetic engineering technologies, dog breeders could repair a disease-causing mutation without sacrificing breed-specific traits, and cattle breeders could move disease resistance between breeds without also introducing phenotypes that are locally maladaptive. Synthetic biology also expands our experimental horizons. We can move traits that evolved in one species into a different species—one with which it would never naturally mate. We can even augment species with traits that are entirely human engineered.

Genetic engineering technologies have existed for nearly as long as artificial insemination and embryo transfer, and I will discuss them in detail in the second half of the book. First, though, let's return to the early twentieth century, when our ancestors' success as engineers of domesticated plants and animals was crippling human society and threatening global habitats with deforestation, overgrazing, and pollution. It was in the midst of this next catastrophe that an idea was born. Perhaps the best way to preserve what wild spaces remained was to make these spaces less wild.

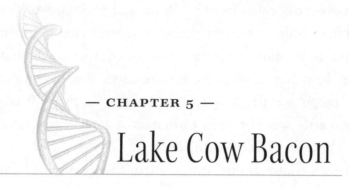

— CHAPTER 5 —

Lake Cow Bacon

[H. R. 23261, 61st Congress, Second Session]
A BILL To import wild and domestic
animals into the United States.

Be it enacted by the Senate and House of Representatives of the United States of America in Congress assembled, That the Secretary of Agriculture be, and he is hereby, directed to investigate and import into the United States wild and domestic animals whose habitat is similar to government reservations and lands at present unoccupied and unused: *Provided,* That, in his judgment, said animals will thrive and propagate and prove useful either as food or as beasts of burden; and that two hundred and fifty thousand dollars, or as much thereof as may be necessary, is hereby appropriated, out of any moneys in the Treasury not otherwise appropriated, for this purpose.

B Y LATE MARCH 1910, THE PEOPLE OF THE UNITED STATES had settled into a collective restless despair. The population had tripled to nearly 100 million over the previous half century.

Forests had been felled and mountains stripped away, exposing the edges of the now-measurable frontier. Bison and passenger pigeons were reduced from billions to near extinction, and markets continued to demand more: more pelts for coats, more feathers for hats, and more beef. Ranchers tried to raise more cattle, but overgrazing destroyed the range and the industry was collapsing. Every season, millions fewer cattle were brought to market. People were hungry and anxious. There was talk of eating dogs.

In the swampy southeastern states, where the quality of forage was too poor to support cattle, another disaster was looming. The Japanese delegation of the 1884 World's Fair had brought as a gift to the host city of New Orleans a water hyacinth, *Eichornia cassipes*. Captivated by the contrast between its tiny lavender flowers and the thick green leaves in which they grew, the people of New Orleans eagerly introduced the water hyacinth into the city's parks and ponds and their backyard streams. Then it began to spread, the patches of green as much as doubling in size every week. By 1910, impenetrable mats of water hyacinths were choking lakes, rivers, and bayous, sucking oxygen from the water and killing fish, and blocking river transportation into the Gulf of Mexico. The Department of War tried clearing it out by hand, poisoning it, even drowning it, but nothing worked. The simple but charming water hyacinth had sown an environmental and economic emergency.

Louisiana's United States congressman Robert "Cousin Bob" Broussard had a plan to solve both crises. He would convince Congress to pass House Resolution 23261, which would provide $250,000—around $6.5 million in today's dollars—to import hippopotamuses into Louisiana's bayous. The hippos would eat the hyacinth, clearing the waterways while transforming the offending plant into tons of delicious hippopotamus meat.

The idea to import African animals into the United States was not Broussard's originally. Four years earlier, Major Frederick Russell Burnham, a scout and adventurer who was the inspiration for the Boy Scouts of America, proposed to his friends in the political elite that the United States should import antelopes, giraffes, and other African animals to populate newly established land reserves in the American West. Burnham had spent much of the previous decades in southern Africa, where his firsthand experience convinced him that bringing these magnificent and edible beasts to the United States was simple common sense. It was clear to him that these animals would not only answer the Meat Question but also foster support, in particular from sport hunting enthusiasts, for the conservation movement. This is the idea that became H.R. 23261.

To make the case for H.R. 23261, Broussard and Burnham enlisted two other unlikely allies. William Newton Irwin from the US Department of Agriculture's Bureau of Plant Industry was officially in charge of improving the country's fruit orchards but was obsessed with solving the meat crisis. Irwin was convinced that expanding ranching in the West, where the land was either occupied or overgrazed, would fail and was instead looking for new places to grow meat. When he heard of the hippo plan, he thought it was perfect: hippos are native to swampy places (in sub-Saharan Africa), they spend their days in the water (out of the way) and graze at night, and they consume 40 kilogram (88 lb) of grass per hippo per day. A hippopotamus, Irwin wagered, could eat a lot of water hyacinths and feed a lot of hungry people. Sold, he joined the team.

The fourth member of their team was Frederick "Black Panther" Duquesne. Like Burnham, Duquesne was an accomplished scout— in fact the two men had been employed to assassinate each other during the Second Boer War. But Duquesne was also a con artist who at other times had worked as a pimp, a German spy, a photographer, a botanist, a fake paraplegic, and a traveling showman. It was while Duquesne was posing in a traveling one-man show as "Captain Fritz Duquesne, adept and legendary hunter of African game" that Broussard stumbled upon him. Duquesne's performance, of which parts were probably true, was sufficient to convince Broussard to invite Duquesne to join his panel of experts.

When Broussard's team assembled to present evidence to Congress in support of H.R. 23261, the committee asked the obvious questions: Are hippopotamuses dangerous? Will they eat water hyacinth? Can they be domesticated? How quickly will their populations grow, and could they become pests? The men answered as best they could. Irwin cautioned that hippos could be destructive if turned loose. Duquesne countered, claiming, based on no evidence, that hippos were naturally tame, could be fed milk from a baby's bottle, and enjoyed being led around on a leash. All three pontificated about the taste of hippo meat (a cross between beef and sweet pork) and about why hippo meat wasn't already a dietary staple (because no one had yet told us to eat hippo meat). From the tone of questioning, it was clear that nearly everyone present believed that hippos could indeed answer the Meat Question and solve the hyacinth problem.

As news of the testimony emerged from Washington, newspapers across the country praised Broussard's creativity and declared enthusiastic support for H.R. 23261. The *New York Times* took to referring to the hypothetical hippos as "lake cow bacon." A *Times* editorial published in April 1910 gushed that "on 6,400,000 now useless acres in the Gulf States, 1,000,000 tons of the most delicious of flesh foods, worth $100,000,000, may be grown yearly." Everything was on track for Broussard's plan to succeed.

The resolution, however, never saw a congressional vote. Burnham, Irwin, and Duquesne testified too late in the year for Congress to act before the end of the 1910 session. Broussard intended to reintroduce the bill, so to continue to raise awareness, the group formed the New Food Supply Society and Burnham planned a research trip to Africa. Then Burnham was called to Mexico to protect copper mines threatened by the beginning of the Mexican Revolution, canceling the Africa trip. Irwin died, the New Food Supply Society fell into disarray, and Duquesne fell into paranoia that others were stealing *his* idea to import hippos. Broussard died in 1918, never having brought the bill back up for discussion.

It's hard to say whether the plan would have worked. Hippos do eat a lot of plants, but they are also highly territorial and dangerous. Hippo meat is edible, but whether people would choose to eat it is uncertain. Whether hippos would be amenable to life in captivity is also questionable. And if the hippo population grew to be large enough to feed a nation of meat-eaters, then they would certainly produce a lot of waste, which would have to go somewhere.

Some answers come from an experiment of sorts taking place today in Colombia, where a herd of hippos lives on and around the estate of the late Colombian drug lord Pablo Escobar. The experiment began when Escobar brought four hippos to his estate in the 1980s. When he died, they were deemed too difficult to remove and left to their own devices. Today, the population has grown to more than 50 hippos, proving that hippos can survive and reproduce on their own outside Africa. Their rapid population growth has also sparked concerns that they might displace native species, which confirms their invasive potential. They are also dangerous. In 2009, three hippos escaped, killing several cattle before one hippo was killed and the others rounded up and returned to Escobar's estate. As to their tamability and delectability, they don't seem particularly fond of humans (casting doubt on Duquesne's testimony), and none are harvested for food. Scientists also disagree about whether their impact on the environment is good or bad. On one hand, they transfer nutrients from the land to the water and create channels through which water flows across the wetlands, services to the ecosystem that have been missing since Colombia's native megafauna went extinct. On the other hand, hippos did not evolve in Colombia and their introduction will alter the evolutionary paths of the species that live there. The hippos do, however, attract tourists, which brings an economic boost to the community.

The popularity and impracticality of H.R. 23261 captures the early twentieth-century clash between our desire to have more now—to take more of the world's resources and turn them into an easier life for more of us—and our realization that natural resources are finite. Broussard's bill was ostensibly about conservation, but its true intent was to conserve a shipping route and a way of life. Any animal that could provide this service, even an enormous and probably dangerous animal that most people had never heard of, was destined to increase the nation's prosperity. No acknowledgment was given to the fact that the introduction of this non-native species was intended to correct the harms of a previously introduced non-native species. No consideration was given to the ecological impact of hippos in the bayous, or to the long-term sustainability of growing, harvesting, and distributing hippo meat. The bill was backed by raw excitement about a solution to the nation's immediate woes.

Despite what seem like obvious flaws, Broussard's hippo bill was ideologically aligned with Progressive-era conservationists. Disgusted by declines of bison and other large animals, sports hunters and landowners had begun to organize against overhunting and deforestation. They wanted to impose limits on taking trees and animals to give populations a chance to rebound, but they knew that this would only be possible if they could get the regular person on their side. They imagined creating publicly owned parks where people could vacation to escape calamitous cities. In these parks, hunting and logging would be allowed, but only at sustainable rates. And the parks could be populated with wild animals—imports from other continents as well as native wild animals—that people would want to see. Progressive-era conservationists saw value in wild animals, but that value was measured as the services these animals provided to people. This human-centric view may seem naïve from today's more ecosystem-focused perspective, but it was exactly what was needed to jump-start land protection and biodiversity conservation. For some species, this happened just in time.

THE END OF PLENTY

When Europeans landed in the Americas in the fifteenth century, mammoths, mastodons, and giant sloths were long gone. Vegetation once kept at bay by these ice age giants was now managed by people who used fire and other technologies to create habitable land and promote the growth of edible plants. Massive bison herds ranged across the North American prairies, as did wild turkeys, wolves, and coyotes. Toward the coasts, abundant forests were being slowly displaced by farms cultivating domesticated plants like corn, beans, squash, and sunflowers.

And there were the pigeons.

Passenger pigeons had long been part of the North American ecosystem. The ancestors of passenger pigeons diverged from those of their closest living relatives, band-tailed pigeons, sometime before 10 million years ago. Passenger pigeons and band-tailed pigeons survived the ups and downs of the ice ages on opposite sides of the Rocky Mountains, with passenger pigeons to the east and band-tailed pigeons to the west. Their diets included the seeds of masting plants: acorns, beechnuts, small chestnuts, pretty much anything they could find. The two pigeon species were ecologically similar except for one notable difference: passenger pigeon populations were enormous.

When Europeans arrived in North America, they encountered flocks of literally billions of passenger pigeons. In 1534, Jacques Cartier became the first European to see a passenger pigeon. Cartier was exploring the eastern coast of Canada when he spotted on what is today Prince Edward Island "an infinite number of wood pigeons."* The passing of passenger pigeon flocks above a settlement would take days to complete, during which "the rays of the sun were

---

* This quote is attributed to Cartier in A. W. Shorger's 1955 manifesto "The Passenger Pigeon: Its History and Extinction," which is a must-read for anyone who is even marginally obsessed with passenger pigeons.

intercepted and daylight was reduced to the dusk of a partial eclipse" and "the dung fell in spots not unlike melting flakes of snow."*

Passenger pigeons were an ecological force with which to be reckoned. Their flocks moved from forest stand to forest stand, consuming everything in their path. Where they nested, each tree might contain 500 nests. When they abandoned the nests, which they did all at once at the end of the nesting season, they left a thick layer of dung on the forest floor and trees denuded and broken. They suffocated the forests, and in doing so reset the ecological cycle, making way for primary forests to move in where mature forests had been.

Embarrassingly, I did not discover passenger pigeons until I was a graduate student. My first ancient DNA research project aimed to figure out whether the dodo was a type of pigeon, as some people suggested, or a separate lineage closely related to pigeons. To answer this I needed dodo DNA, and I was hoping to get permission to cut a tiny bone sample from the Oxford University Museum of Natural History's famous dodo for DNA analysis. The zoology collections manager, Malgosia Nowak-Kemp, was a slight but fiery woman who took tremendous pride in each of the museum's specimens. She was anxious to see the specimens used for science but also nervous about the idea of putting holes in them. She wanted to help me with my research but, before she would give me permission to cut into the dodo, I would have to prove my chops, as it were, on less precious dead birds.

Keen to show off her collection of extinct pigeons, Malgosia dragged me and Ian Barnes (who always seems to accompany me on museum misadventures) up a flight of stone stairs and into the foyer of a grand storage room. Rows of floor-to-ceiling metal cabinets filled the room, identical apart from the writing on small white cards taped to their doors that bore the taxonomic names of the

---

* John J. Audubon includes this delightful recollection of his time spent observing passenger pigeons in Ohio in his entry about passenger pigeons in his *Ornithological Biography*, 1831–1839.

dead things within: *Galliformes, Anseriformes, Psittaciformes* . . . She instructed us to wait by the door and disappeared into the room, searching for the cabinets of dead pigeons, *Columbiformes*. The next minutes were filled with the sounds of stomping, dragging drawers out of cabinets, opening and slamming doors, and sporadic cursing in a combination of English and Polish. Ian and I waited uncomfortably, not sure whether to offer to help or stay out of her way as we had been instructed. Finally, she emerged triumphant, a stack of paperwork in one hand and a small, mounted display in the other in which a brightly colored passenger pigeon stood on a thin American chestnut branch. Threading her way toward us through the cabinets, Malgosia's eyes sparkled with a combination of motherly pride and admiration. Her voice cracked with emotion as she explained that this display had been in the main hall but was returned to the storeroom while the exhibit underwent an upgrade and that I could, if I wanted, have a small piece of its toepad for my DNA study.

What happened next is burned into my memory. The stone floor in the storage room was uneven and oddly sloped, which was not unusual for a building that had been in use since the 1850s. As she walked toward us, her eyes and mind focused on the pigeon, Malgosia misjudged a step around a particularly unsettled stone. She recovered quickly, turning what might have been a painful fall into a light stomp. But as her foot hit the ground, the display shook, and all three of us held our breath. The pigeon wobbled, and then settled down. As we exhaled, we heard a light "pop," as the branch on which the pigeon was perched snapped. We watched in horror as the bird fell forward initially, then hit the base of the display with its head, knocking its feet into the air. Time slowed as the pigeon flipped toward the ground, completing at least one 360-degree rotation before landing on its back with a dull "thunk." Ian and I stared at the newly dismounted passenger pigeon, too nervous to look up. But it wasn't over. A second or so after landing, the pigeon's head separated neatly from its body and, as a final indignity, began to roll away, with a faint "click, click, click" resonating down the rows of cabinets each time its beak hit the stone floor.

I, of course, did what every self-respecting early career, ancient-DNA scientist would do. I took a piece of its toepad and extracted its DNA.

We didn't learn much about the passenger pigeon from this first genetic analysis. I amplified a few small fragments of DNA and compared these to sequences from other pigeons, including the Oxford dodo that I eventually earned permission to sample (and which we now know is a type of pigeon). These small fragments of DNA linked passenger pigeons to the group that includes band-tailed pigeons, but told us nothing about why passenger pigeon flocks were so unusually large.

How does a population decline from billions to none in only a few decades? Unfortunately for passenger pigeons, they were as tasty as they were abundant, and they were easy to catch and kill. Even small children could hunt them successfully by poking birds out of trees with sticks or knocking them out of the sky with potatoes. A decent throwing arm, however, was not the real threat to passenger pigeons. The birds were instead doomed by technology. During the first half of the nineteenth century, European colonists laid an expansive network of railroad tracks across the continent. This network created a high-speed connection—faster than the flocks could fly—from one end of the passenger pigeon range to the other, and from every potential passenger pigeon nesting site to the hungry markets of the East Coast. By 1861, intercontinental telegraph lines made it possible to signal the location of passenger pigeon flocks to whoever wanted to board that train. Within the next two decades, billions of passenger pigeons were located, tracked, and killed. They were netted, shot, poisoned, and suffocated by placing pots of burning sulfur under nest-filled trees. Individual hunters took as many as 5,000 birds each day. In 1878, near Petoskey, Michigan, and at what would be the last known large passenger pigeon nesting site, 50,000 birds were killed every day for almost five months. By 1890, the entire wild population included only a few thousand birds. In 1895, conservationists collected the last wild passenger pigeon nests and eggs to try to establish captive populations. In 1902, the last wild passenger pigeon was shot by hunters in Indiana, and Martha, one of the few remaining wild-born passenger pigeons, was sent to the Cincinnati Zoo and put in a cage with George, who had been born in captivity. Martha and George never bred, and George died in 1910. Martha lived alone, as the last of her kind, for just over four years. She and her species died on the first of September in 1914.

Even with the combined misfortunes of the rail and telegraph systems and the ease with which passenger pigeons could be caught, the decline of passenger pigeons was curiously quick. One hypothesis to explain why no small populations persisted posits that the large flocks were actually necessary for their survival. This phenomenon is known as the Allee effect, named after ecologist Warder Clyde Allee. Allee discovered in the 1930s that goldfish grew more rapidly when they lived in confined quarters with other goldfish than they did when they lived alone. This was contrary to expectations; competition for finite resources should make individuals grow more slowly rather than more quickly. Allee proposed that the answer was cooperation. If individuals cooperate, for example to find food or to defend against predators, then more individuals would mean more cooperation and increased individual fitness. Translated to passenger pigeons, the Allee effect would mean that, over time, passenger pigeons may have evolved to rely on cooperation, and therefore large populations, for their survival.

To evolve an ecological strategy in which large populations are required for survival, passenger pigeon populations would have had to be very large and for a very long time. When we first began to study them, most people believed that passenger pigeon flocks had only become large recently, either as forests expanded after the last ice age or after the first Americans began to practice agriculture at scale, as crops would have made ample food sources for growing pigeon populations.

In 2000, when I extracted mitochondrial DNA from Oxford's headless passenger pigeon,* DNA sequencing technology was not yet sufficiently advanced to allow us to generate whole genomes for extinct species. This changed during the first decade of the twenty-first century, as new DNA sequencing technologies allowed access to even the tiniest surviving fragments of ancient DNA, and, with these tiny fragments, to piece together entire genomes. Whole genomes would provide much greater power to determine when passenger pigeon populations first became large and perhaps why they went extinct as quickly as they did.

Over the next few years, we isolated DNA from hundreds of passenger pigeon skins, feathers, and bones, searching for a specimen that was sufficiently well-preserved to generate a complete genome sequence. Finally, in 2010, Allan Baker, then the head of the department of natural history of the Royal Ontario Museum, granted us permission to sample 72 (!) passenger pigeons from the museum's collection, which included some of the last passenger pigeons to be shot and preserved. Among these, we found three that were extremely well-preserved. Focusing on these, we generated billions of short DNA sequences and carefully assembled them into genomes. We then compared these genomes to each other and to genomes from band-tailed pigeons. We estimated when passenger pigeon populations became large and looked for genetic mutations that may have made them more successful in large populations.

Our results were both satisfying and not. We learned that passenger pigeon populations have been large, very large, for at least the last 50,000 years and probably for many tens of thousands of years longer. This means that their flocks were giant even during the

---

* Actually, Malgosia glued the head back on after this unfortunate incident, and the specimen is now completely intact.

coldest part of the last ice age. We found genetic changes that affected their responses to stress and helped them fight diseases, both situations that one might expect in extremely large and dense populations. We found genetic hints that passenger pigeons were dietary generalists, which they must have been to survive through such different climatic regimes. But we did not find any genetic clues that might reveal why they went extinct.

It is possible that, as passenger pigeons evolved in large flocks, they lost behaviors that would have been important in a solitary life. There's little need to be really good at finding food or a mate, for example, if there are millions of individuals in your immediate proximity with which you can cooperate. This is only speculation, however; there is as yet no genetic evidence supporting an Allee effect contributing to passenger pigeon extinction, and the limited data that we have suggest that captive birds reproduced just as readily in groups of a dozen or so as they did in the wild in groups of several hundred million. It seems most likely that passenger pigeons went extinct because people killed them. Our interventions were too little and too late.

While passenger pigeons are gone forever, their decline did not go unnoticed. The perception of their imminent loss, which was occurring simultaneously with that of American bison, set in motion some of the earliest legal efforts to protect endangered species. Their extinction paved the way for many of the conservation laws, protocols, and regulations that are in place around the world today.

A GIFT TO THE AS YET UNBORN

In 1842, the United States Supreme Court ruled in *Martin v. Waddell's Lessee* that a property owner did not have the right to exclude others from harvesting oysters from a mudflat off the coast of his property in Raritan Bay, New Jersey. This decision meant that the court considered land beneath navigable waters the property of the public rather than of any private individual. This *public trust doctrine*—the idea that wildlife and wildlands belong to everyone and are the government's responsibility to protect— became the cornerstone of the North American Model of Wildlife Conservation.

At the time of the *Martin v. Waddell's Lessee* decision, North American wildlife was in trouble. European settlers had cleared the forests along the eastern coast of the continent, turning the trees into ships and planting crops where the trees had stood. Every year these ships carried tens of thousands of animal pelts—beaver, deer, bison, and others—to England, creating and then satisfying market demand. By the middle of the seventeenth century, beavers had been nearly eliminated from the eastern coast and deer were so rare that the Massachusetts Bay Colony offered to pay one shilling for each wolf killed—assuming implicitly that the decline of deer was because there were too many wolves rather than too many hunters.

Environmental destruction was not limited to the East. Along the coasts of the northern Pacific, the Russian-American Fur Company was specializing in sea lions, Steller's sea cows, sea otters, and northern fur seals. The scale of the killing was impressive. Several years ago, while looking for mammoth bones on St. Paul Island in the Bering Sea, I stumbled across what I thought was evidence of an ancient tsunami. A storm the night before had exposed a wall of sand along the coast, revealing strata that accumulated over the last several hundred years. A few meters from the top I noticed several small bones protruding from the sand and began to dig them out. It wasn't until I had collected about a dozen fur seal skulls, each with a round, tennis-ball-sized indentation, that I knew what I'd found. This had been a northern fur seal rookery and was now the site of a several-hundred-year-old massacre.

The consequences of these targeted hunts varied, with some species like sea otters narrowly avoiding extinction, and others like Steller's sea cows disappearing entirely. Steller's sea cows were

10-metric-ton (22,000 lb), 9-meter-long (30-ft-long) relatives of dugongs and manatees that lived in the underwater kelp forests that once thrived along the coasts of the northern Pacific Ocean and Bering Sea. Steller's sea cows were hunted for their flesh and their fat, which people ate and burned to stay warm while hunting for smaller marine mammals. They went extinct within 27 years of their first reported sighting by Europeans. Jim Estes, an ecologist who has spent much of his career studying the giant kelp ecosystems, believes that sea cows were doomed even if people hadn't found them palatable. Their fate was sealed, he argues, when people extirpated sea otters. With no otters to eat sea urchins, urchin populations exploded. The urchins consumed and destroyed the kelp forests on which the sea cows relied for food and protection. Exposed and vulnerable, the gentle giants didn't stand a chance. Their story highlights the delicate balance of ecological communities, in which every species relies on other species for survival. When humans disrupt this balance, the entire ecosystem is affected.

As land and animals became scarcer in the East, the colonists expanded westward. European diseases had spread centuries earlier, ravaging Native American populations and reducing human predation of native wildlife. As colonists spread into this vast land of plenty, they saw no reason to limit what they took, in particular while markets in Europe and in the cities along the US East Coast were willing to pay for fur and meat. They trapped all the beavers they could find and, when they couldn't find any more beavers, they became bison hunters. By the early 1800s, a system of trading posts had emerged across the west that encouraged Native Americans, whose bison hunting skills were augmented by the reintroduction of the horse, to take more animals and sell their hides. Bison were slaughtered by the thousands, sometimes for their pelts, sometimes to feed crews building railroads, and sometimes just for their tongues, which were considered a delicacy.

In the West, neither Native Americans nor the early colonists were in a position to stop the destruction. Native Americans were being pushed slowly and callously from their land onto reservations and were fighting for survival. Colonists saw wildlife as one of their only sources of income and sustenance. In the East, however, change was brewing. As people became wealthier, they started to have time for recreational activities, and sport hunting was a favorite. Sport hunting, however, requires that animals exist. It should be no surprise then that sportsmen's clubs played an important role in establishing animal protection laws.

Among the most influential of these clubs was the New York Sportsmen's Club, which was founded in 1844 to enforce the public trust doctrine established by the Supreme Court ruling a few years earlier. The New York Sportsmen's Club drafted game laws, campaigned for enforced closed seasons, hired informants to find people violating wildlife and game laws, and sued violators using members' own money and legal expertise.

While the New York Sportsmen's Club was busy championing game animals as a public trust, other clubs were campaigning for wild places to be established as public land. The clubs' efforts paid off. In 1864, President Abraham Lincoln signed the Yosemite Grant Act, granting 39,000 acres of Yosemite Valley and the Mariposa Big Tree Grove to the state of California and protecting the valley from commercial development. In 1872, President Ulysses S. Grant signed the Yellowstone National Park Protection Act, establishing Yellowstone as the first national park in the United States and protecting it, too, from private development. Although the army had to be used for more than a decade to keep squatters and poachers out of the parks, the creation of these public lands was immediately impactful. For example, most bison alive today trace their ancestry back to the population that lived within the protected borders of Yellowstone National Park.

This was the backdrop to the 1878 passenger pigeon slaughter at Petoskey, Michigan. Many states had laws by this time purporting to protect game animals, but these laws were weak and difficult to enforce, and the culture around hunting was still driven by market forces. As the date of the planned slaughter approached, small teams opposed to the pigeon kill traveled to Petoskey to do what they could to thwart it. They smashed a few traps and coerced local authorities to distribute a fine, but they had little impact against the thousands who turned up to kill birds. In the end, nearly a billion passenger pigeons were killed in under two months in Petoskey, and the last nesting colony of passenger pigeons was no more.

Michigan passed a law in 1898 that banned the killing of passenger pigeons for a period of 10 years, which officials deemed would be enough time for the birds to recover. But it was too late. Passenger pigeons were already functionally extinct in the wild. Bison, too, were on their way out, having missed an opportunity for legislative protection when President Grant vetoed a bill that would have made it unlawful to kill female bison. It was not all bad news, however. The near disappearance of these two once numerous animals was widely reported in the media, and people began to mourn them. Not just their flesh, but the animals themselves, their giant flocks and herds, and the colonial spirit that these flocks and herds represented. The declines of the passenger pigeon and the American bison became rallying cries for the conservation movement.

President Theodore Roosevelt did more than any American president to make conservation a priority in the United States. Teddy Roosevelt detested the decline of the bison, having seen it firsthand at his Elkhorn Ranch in North Dakota. When he became president in 1901, he began a series of sweeping changes that laid the foundation for today's conservation laws. With conservationists like Gifford Pinchot and preservationists like John Muir as advisors, he helped to establish more than 230 million acres of public lands during his presidency. He created the United States Forest Service and the Bureau of Biological Survey; the latter would merge several decades later with the Bureau of Fisheries to become the United States Fish and Wildlife Service.

Soon, good intentions started to sync with legislation and science. In 1900, Congressman John F. Lacey, a Republican from Iowa, introduced a wildlife protection law now known as the Lacey Act that banned interstate shipping of game that had been killed illegally. The Lacey Act also authorized the government to restore wildlife where populations had disappeared or were threatened with disappearing. In 1905, conservationists founded the American Bison Society, with Teddy Roosevelt as its honorary president, and set into motion a plan that would eventually save bison from extinction. During the first two decades of the twentieth century, the sciences of wildlife and forest management began to move away from taxonomic stamp collecting to evidence-based consideration of how organisms fit into ecosystems. The Ecological Society of America and American Society of Mammalogists were established, and both created scientific journals to help communicate data and ideas. Scientists began to develop methods to census populations, to understand plant succession, and to appreciate community interconnectedness through food webs. People began to consider the impact of extinctions not just on the economy but also on ecosystems.

By 1910, every state had some sort of commission to protect wildlife, but few had means to pay for that protection. This changed when Pennsylvania passed a law requiring hunters to purchase a license to take game legally. When other states saw that Pennsylvania suddenly had money to support game-law enforcement and wildlife restoration, they, too, began to charge for hunting licenses. In 1937, the Pittman-Robertson Federal Aid in Wildlife Restoration Act added an 11 percent excise tax on hunting supplies— weapons and ammunition—with revenues to be distributed to state wildlife agencies for conservation and hunter education programs. Wildlife conservation was growing teeth.

Predictably, not everyone was on board with paying for hunting game. Some opponents argued that hunting could not be limited legally because the wildlife itself was owned by the people—with reference to the public trust doctrine. In 1916, Teddy Roosevelt wrote in response, "So it does; and not merely to the people now alive, but to the unborn people. The 'greatest good for the greatest number,' applies to the number within the womb of time, compared to which those now alive form but an insignificant fraction."

AN ANSWER TO THE MEAT QUESTION

When Cousin Bob Broussard died in 1918, the world was embroiled in the Great War, and both the Meat Question and the hyacinth problem had disappeared from the front pages of the nation's newspapers. The Department of Agriculture had given up on hippos and was instead encouraging Louisiana to transform its swamps into pastures by reclaiming the land—essentially creating mud barriers to block water from entering so that marshes would dry up and grow cattle-friendly food. Technological innovations—machines that made the harvest faster and more efficient, and chemicals that eliminated pests and increased the land's fertility—were also making it possible to do more with less. It was no longer necessary to find new land or diversify the menu. People could simply grow more cattle in the space they already had.

Some technologies had unpleasant side effects. The copper arsenite insecticide Paris green was first dusted onto potato plant leaves in the 1860s to limit crop destruction by Colorado potato beetles. By the turn of the twentieth century, farmers had used so much of the insecticide that officials found it necessary to pass legislation to control the use of chemicals in agriculture. Even the new machines were causing trouble. When California's

Holt Manufacturing Company introduced a self-propelled combine harvester in 1911, farmwork that had been done by hand suddenly didn't need as much human input. Small, diversified family farms were replaced by larger and more efficient farms specializing in monoculture. Farm consolidation continued throughout the 1920s, through the Great Depression, and into the middle of the twentieth century. By the 1940s and 1950s, the same amount of land was supporting many times the number of animals than it had decades earlier, creating ideal conditions for disease. During the 1940s, scientists discovered that adding antibiotics to animal feed both protected them and caused them to gain weight faster. This practice of preventative drug treatment for farm animals continues today, with the World Health Organization reporting in 2017 that in some countries as much as 80 percent of available antibiotics are used to make healthy animals grow more quickly.

The consolidation and industrialization of agriculture created new conflicts between landowners and wildlife. Although public support in the United States for conservation had grown during the first half of the twentieth century, government backing of conservation initiatives waned during the prosperous 1950s. The newly wealthy public flocked to national reserves and parks for recreation, encouraged by the automobile industry ("See the USA in your Chevrolet!"). The government, focused on building the military, began to grant licenses to graze cattle, log for building materials, and explore for oil on public lands. The population boomed, and the agricultural industry once again struggled to keep up with demand. New technologies like artificial insemination improved the industry's efficiency, but crop and livestock populations were at capacity. Eventually, more land that had been wild was converted for commercial housing, industrial plants, and agriculture. Roads, dams, ploughed and pesticide-filled fields, and the ever expanding human footprint continued to fragment the spaces that remained.

In June 1962, the *New York Times Magazine* published an excerpt
from Rachel Carson's new book, *Silent Spring*, in which Carson describes a desolate future where our indiscriminate use of pesticides
has caused, among other horrors, the death of all birds and complete loss of birdsong. Her message was chilling and intended to
be so. She compared pesticides explicitly to nuclear fallout, calling
both invisible and inescapable threats, and accused the chemical
industry of colluding with governments to spread disinformation
about the dangers of pesticides. The industry fought back, branding
Carson a communist, an amateur, and hysterical, and threatening
her publisher with a libel suit even before the book was released.
President John F. Kennedy, however, convened an official panel to
investigate her claims, eventually leading to changes in how chemical pesticides are regulated in the United States. This was not *Silent
Spring*'s only legacy. Carson's book incited a grassroots environmental movement that eventually led to a global ban of the pesticide
dichloro-diphenyl-trichloroethane, or DDT, helped to create the
Environmental Protection Agency in the United States, and continues to inspire environmental movements around the world.

THE CONSEQUENCES OF ACTING

On March 11, 1967, as the Endangered Species Preservation Act
of 1966 went into effect, 78 species native to the United States
collectively became the first officially protected endangered species. The "class of 1967," as the US Fish and Wildlife Service
calls them, included grizzly bears, bald eagles, American alligators, black-footed ferrets, Apache trout, Florida panthers, California condors, and whooping cranes, the iconic, tall, snowy-white
bird with a crimson cap whose decline to only 43 individuals had
helped to inspire the legislation. Each of the 78 listed species was
considered extremely endangered. With the passing of the act,

people took formal control over their future, completing our transition into the role of protector. From that point forward, people would decide whether and how each of these 78 species survived.

The Endangered Species Preservation Act instructed federal agencies to establish plans for each listed species' recovery and provided funds to make their recovery possible. In 1967, conservation scientists established a whooping crane breeding colony at Patuxent Wildlife Research Center in Maryland with three eggs collected from nests in Canada's Wood Buffalo National Park. Eight years later, they placed whooping crane eggs sourced from the captive breeding colony into the nests of sandhill cranes at Gray's Lake National Wildlife Refuge in Idaho. The sandhill cranes raised the whooping crane hatchlings as their own and, fifty years later, more than 700 whooping cranes were recorded in the wild and in captivity in North America. For fifty years wildlife managers decided which birds bred, fed them, protected them, and released them into ideal habitat. By manipulating the birds' evolutionary trajectory, these wildlife managers prevented their extinction. And whooping cranes are not the only species saved thanks to human effort; in February 2020 the Hawaiian ('Io) Hawk became the seventh of the class of 1967 species to be formally removed from the endangered species list, joining the Canada goose, American alligator, Delmarva Peninsula fox squirrel, longjaw cisco, Mexican duck, and bald eagle as conservation success stories.

Endangered species protection grew in reach and potency throughout the second half of the twentieth century. In 1969, the US Congress amended the Endangered Species Preservation Act to prohibit importing or selling at-risk species from other parts of the world. In 1973, 80 countries signed the Convention on International Trade in Endangered Species of Wild Fauna and Flora, or CITES, bolstering the prohibition on international trade of endangered species. Later that year, US president Richard Nixon signed into law a completely overhauled Endangered Species Act, or ESA.

The ESA enforced CITES rules in the United States and added invertebrates and plants to the list of species that could be declared threatened or endangered. In 1993, the United Nations Environment Programme enacted the Convention on Biological Diversity. The program, which recognizes that biodiversity is integral to global prosperity and that coordinated international collaboration is critical to conservation success, has near-global support.

PANTHER CROSSING

While the ESA provides a legal framework to protect threatened and endangered species in the United States, it is far from perfect. Every listed species requires its own recovery plan, which often must be developed with limited knowledge of the species' evolutionary history or habitat requirements. Federal agencies must avoid doing anything that might jeopardize listed species on their lands. If a listed species is discovered on privately owned land, the landowners may not hunt, shoot, wound, trap, harass, take, harm, or pursue that species. Unsurprisingly, these strict rules lead to hundreds of legal challenges every year that pit the ESA against the rights of property owners. With near-constant legal challenges, there is a tendency among wildlife managers to avoid unnecessary risk and stick with conservation strategies that limit human activities that might negatively impact listed species. Today, however, there is a growing realization that these passive strategies have not slowed the rate of biodiversity loss, and that our role as protectors is to intervene.

One of the highest-profile endangered species in North America was saved because people intervened. Florida panthers are an ecotype of puma, a large cat that goes by many names, including mountain lion, catamount, painter, cougar, and mountain screamer, the latter for the bone-chilling call that females make when searching for a mate. When Europeans arrived in the Americas, pumas were distributed nearly continuously from Canada's Yukon to the southern tip of Chile. By the middle of the twentieth century, however, hunting, deforestation, and agriculture had restricted them to pockets of refugial habitat across this range. When they were listed in 1973, the Florida panther had been reduced to fewer than 20 animals living in the southern tip of the state of Florida.

Fast forward to 1981, when Florida Game and Fish Commission biologist Chris Belden was leading development of the Florida panther recovery plan. To make a plan enforceable, Belden's team needed to identify physical traits that distinguished Florida panthers from other pumas so that managers and members of the public could recognize them as such. Over several years, Belden's team trapped a dozen or so of the reclusive panthers of Big Cypress National Preserve and recorded data describing their health and physique. They noted the panthers' wide flat foreheads and highly arched nasal bones; these were the traits that paleontologist Outram Bangs had used when describing Florida panthers as a distinct subspecies in 1899. Belden's team, however, chose two different traits to define Florida panthers: a cowlick on the back of their neck and a kinked tail.

Cowlicks and kinked tails are unusual traits to select as group-defining. From an evolutionary perspective, these are disfigurements that one might expect to be removed by natural selection in a healthy population rather than evolutionary novelties that are either neutral or beneficial. Of course, the Florida panther population was not healthy. After decades of isolation, their only choices of potential mates were their own siblings, cousins, and parents. Generations of inbreeding reduced their genetic variation and increased the frequency of maladaptive traits like kinked tails. The panthers were also suffering from more serious genetic abnormalities. By the early 1990s, 90 percent of Big Cypress males had a disease called cryptorchidism in which at least one testicle fails to descend, and more than 90 percent of sperm that the males produced was abnormal. They also had a high incidence of heart defects and suppressed immune systems, making them less able to fight off disease.

There was another problem with defining Florida panthers by kinked tails and cowlicks. While all panthers in Big Cypress National Preserve had these traits, most of the panthers that lived farther south in the Everglades lacked the kinked tail and cowlick. If these were defining traits of Florida panthers, should the Everglades population be considered something else?

Imagining that DNA could solve their taxonomic troubles, evolutionary biologist Steve O'Brien and Melody Roelke, the recovery team's veterinarian, set out to compare the Everglades panthers' DNA to that from other pumas. Their results ground Florida panther recovery to a near halt. The Everglades panthers, it turned out, were not entirely Floridian. Sometime in their recent history, Everglades panthers had exchanged genes with panthers from Costa Rica.

After some digging, Roelke and O'Brien discovered that, a few decades earlier, Everglades National Park rangers had asked Les Piper, the director of a roadside zoo in Bonita Springs, to release several of his captive Florida panthers into the park to augment the park's declining population. Unbeknownst to the rangers, Piper had already augmented his captive population with individuals from Costa Rica. He'd done this to improve their breeding success, and it had worked stunningly well. This gave them an idea. The relative success of the outbred Everglades panthers compared to the inbred Big Cypress panthers suggested that Florida panthers could be rescued by introducing variation from another population.

Unfortunately, the US Fish and Wildlife Service had an unwritten rule against protecting hybrids, which these outbred Florida panthers would be. They feared that protecting hybrids would pollute the gene pool of the endangered species and reduce their defining characteristics (even if these defining characteristics were traits like crooked tails and cowlicks). Some members of the recovery team were so concerned that they asked O'Brien and Roelke to keep their results secret, worried that the mere existence of hybrids would endanger the Florida panthers' protected status. As dissonance grew among the recovery team, the situation in Florida began to spiral out of control. Rumors circulated that panthers in southern Florida were not actually Florida panthers and could therefore be killed. A legal case involving taking (and eating) a panther was dismissed on taxonomic grounds, with lawyers for the defense arguing that if experts couldn't agree about what a Florida panther was, then a layperson certainly shouldn't be expected to do so. And the Florida panther population continued to decline.

Steve O'Brien told me that he was convinced at the time that the only way to save Florida panthers was to convince the US Fish and Wildlife Service to change their minds about hybrids, and he set out to do so. He teamed up with evolutionary biologist Ernst Mayr, the scientist who had first theorized the Biological Species Concept in 1940 and someone who O'Brien hoped would have the gravitas to capture regulators' attention. Together, they wrote an opinion piece for the journal *Science* in which they pointed out that hybridization occurs naturally between subspecies, which by definition are the same "species." They noted that hybrid zones—places where related groups overlap in range and interbreed—are common and yet do not cause the separate forms to blend together into a single form. And they proposed a new definition for subspecies as entities that have a unique natural history, live in distinct habitats or ranges, and share heritable and defining molecular or physical traits. At the core of their argument was that subspecies, even if they mix with other species every now and then, retain recognizable distinctiveness that is sufficient for a wildlife manager or layperson to recognize.

The Fish and Wildlife Service was persuaded and agreed to allow eight healthy female Texas panthers, the geographically and evolutionarily closest population to Florida panthers, to be captured and introduced into the Big Cypress Swamp. Four years later, the first generation of hybrid kittens was born. Fifteen survived, as did six of the introduced Texas panther females. More hybrid kittens were born over the next several years. The hybrids were strong and healthy. The incidences of debilitating disease and physical deformity declined, and the Florida panther population grew by more than 50 percent.

The abandonment of the hybrid policy, which happened without any official change in the rules, benefited other species as well. Barred owls expanding westward across North America sometimes hybridize with threatened northern spotted owls. In 2011, a revision of the northern spotted owl recovery plan noted that hybridization with barred owls occurs but is rare and inconsequential to northern spotted owl recovery. Similarly, natural hybridization between the threatened pallid sturgeon and the more common shovelnose sturgeon in the lower Mississippi and Atchafalaya Rivers is not considered a threat to pallid sturgeon recovery. Hybrids can maintain protection under the ESA as long as their distinguishing characteristics remain intact.

Problematically, hybridization does sometimes blur distinguishing characteristics. The westslope cutthroat trout is found across the Pacific Northwest. For decades, enormous numbers of hatchery-reared rainbow trout have been introduced—dumped by the thousands into lakes and streams from airplanes—into westslope cutthroat trout habitat as stock for sport fishing. These species readily breed, and both rainbow trout and hybrids can outcompete purebred westslope cutthroat trout. Several decades ago, conservation groups petitioned to have westslope cutthroat trout protected under the ESA. When scientists performed their population assessments, however, they were unable to distinguish purebred westslope cutthroat trout from hybrids and couldn't make a recommendation about protection. There is still no solution to this problem, although ancient DNA offers a sliver of hope for the future. In collaboration with Carlos Garza and Devon Pearse from the National Oceanic and Atmospheric Administration's Southwest Fisheries Science Center, my lab is sequencing DNA from trout collected during the early twentieth century—before the airplane dumping began in earnest. We hope to find genetic markers in the museum-preserved trout that distinguish native species from introduced hatchery-reared species, and to use these to develop an approach that wildlife managers can use to prioritize populations for protection. If we're successful, similar approaches may also help guide the US Fish and Wildlife Service to find solutions to other hybrid challenges.

Without any official policy toward hybrids, regulators from the US Fish and Wildlife Service make decisions on a case-by-case basis. Today, genetic data reveal that hybridization is more common than biologists thought it was before the era of DNA sequencing. And the evolutionary consequences of hybridization vary, making it difficult to imagine an effective yet sweeping hybrid policy. Hybridization may have little impact at all, as with northern spotted owls and pallid sturgeon. Or, as with westslope cutthroat trout, hybridization can be destructive, overwriting defining traits and threatening the survival of an endangered species. Or hybridization can be constructive, providing a DNA booster shot of sorts that rescues a population like Florida panthers from the negative effects of inbreeding.

Twenty-five years after US Fish and Wildlife intervened to rescue Florida panthers by hybridizing them with Texas panthers, the population is still healthier than it was in 1995, and more animals are alive than prior to their rescue. Our job, however, is not over.

In 2019, Steve O'Brien collaborated with my research group to sequence DNA from three of the Florida panthers that his team collected during the early 1990s. Two were from the Big Cypress population and one was from the Everglades population, which had unfortunately gone extinct by the time Texas panthers were brought to Florida. We compared each individual's genome to a larger data set of puma genomes from across the species' present range. As expected, the Big Cypress panthers' genomes showed clear signs of inbreeding, and the Everglades panther's genome contained signals of recent hybridization. Parts of the Everglades panther's genome looked like that of Big Cypress panthers, with long stretches of DNA where it was clear that its mother and father were closely related. Other parts of its genome had variation introduced from its Central American ancestors. As we scanned along the Everglades panther's chromosomes, the amount of variation flipped between these two states: lots of variation (the genetically rescued state) and no variation (the inbred state). These two states reveal what happens after genetic rescue. In just a few generations, all the Everglades panthers had long spans of DNA where the variation introduced through outbreeding had been lost. The benefits of hybridization were being overwritten by continued inbreeding.

We have not tested whether the diversity gained when Texas panthers were introduced into Big Cypress is also being lost to inbreeding, but I expect it is. Genetic rescue worked, but the population is still small and isolated, and Florida panthers still have little choice other than to breed with close relatives. Although the population seems healthy, an August 2019 *New York Times* report included videos of several Florida panthers suffering from a neurological disorder that made it difficult for them to control their hind legs. Wildlife managers don't yet know whether this most recent affliction is due to a new genetic mutation or to something else, perhaps a toxin in their environment. What is clear, however, is that their future is our responsibility. We must discover what is causing this latest threat to their survival and find a solution. Otherwise, Florida panthers will become extinct, despite the decades of effort that we have already invested to save them.

Florida panthers would not have recovered without active intervention. People altered their habitat to such an extent that panthers can't get into or out of the population; no amount of passively leaving them alone would have solved the inbreeding depression that led to their poor health or inability to reproduce. Other threatened puma populations might be saved by creating wildlife corridors through which individuals can disperse naturally. Where that is not possible, wildlife managers must replicate this process by physically moving animals between populations. They must also do this with the same regularity as would have occurred naturally, or the intervention will not succeed.

Florida panthers are a hopeful example of what conservation can achieve when people are willing to intervene, but they are also a reminder that human actions have consequences. Florida panthers today are different than they were before, and they are different than they would have been without our intervention. Essentially, people both saved and created what we now know as Florida panthers.

THE HYACINTH PROBLEM

The rate of species extinction today is high, and yet undoubtedly would be even higher had people continued along their nineteenth-century trajectory of exploitation. On every continent, private and public programs have been established to protect wildland from development. Hundreds of grassroots and government-sponsored conservation organizations have been formed with remits that span protecting particular species or ecosystems to ending poaching or whaling to educating businesses and communities about the benefits of biodiversity conservation. Although development hasn't stopped and the human population is still growing, people value biodiversity more than they did in the past, and this makes efforts to protect biodiversity more welcome and more enforceable. Of course, a lot still needs to be done. And while existing approaches to conservation have been effective, they are not enough. Meeting present-day conservation goals will require better and more sophisticated technologies and more willingness to intervene. We need another technological revolution.

Consider the water hyacinth.

Since Cousin Bob Broussard introduced his bill in 1910, the people of Louisiana have tried everything other than hippos to get rid of water hyacinths. They've pulled plants by hand, set them on fire, tried to sink them with oil, and sprayed them with pesticides. When physical and chemical control didn't work, they turned to other species to do the job. They brought in Chinese grass carp that nibbled a few other invasive aquatic plants but found the water hyacinth unpalatable. They imported three insect species—two weevils and a moth—that evolved to specialize in the water hyacinth. The insects did enough damage to make the hyacinth susceptible to disease and suppress flowering, but the water hyacinth continues to spread. Today the hyacinth's thick, green, sunlight-blocking, oxygen-depleting, fish-killing mats clog waterways and wreak economic and ecological havoc on every continent except Antarctica. We need another solution.

Invasive species are not a recent phenomenon. Species disperse naturally, sometimes over long distances. Fish eggs and plant seeds can be carried across continents and oceans by migrating birds. Storms and currents can disperse plants and animals atop floating mats of vegetation. Logan Kistler, an ancient-DNA scientist who specializes in domesticated plants, and I showed several years ago that bottle gourds dispersed from Africa to the Americas by riding transoceanic currents for hundreds of days, their seeds somehow remaining sufficiently viable to establish populations that were later discovered and domesticated by Native Americans.

Once humans invented long-distance travel, they became another vehicle to disperse species, sometimes purposefully and sometimes not. As technologies improved, the pace of human-mediated dispersals quickened. European colonists brought species with them that reminded them of home. Later, people imported species with a particular result in mind. Kudzu, for example, was introduced to the United States from Asia at the end of the nineteenth century to control soil erosion. Kudzu has been so successful in parts of the United States that it's sometimes called "the vine that ate the south," after its propensity to smother vegetation, power lines, road signs, stationary vehicles, and anything else in its path as it grows at a rate of up to 30 centimeters (1 ft) a day.

People are still moving species from place to place that look or taste exotic, that we imagine might be useful in some way, or that we simply don't know we are dispersing. In 2016, an inspector from the US Department of Agriculture seized a package at the San Francisco airport that contained a nest of Asian giant hornets—also known as "murder hornets"—in which pupating insects were very much alive. Asian giant hornets are sometimes eaten as a delicacy or pain remedy, so inspectors assumed that the package was intended to be a gift rather than ecological sabotage. While these inspectors successfully foiled the murder hornet invasion of 2016, adult Asian giant hornets were spotted in both Washington State and British Columbia in 2019, sparking uproar over the impending destruction of native honeybees. With no capacity to observe political borders or human laws, species will continue to move into habitats in which they are considered invasive. If the climate in these habitats suits them, if they can find enough to eat and avoid being eaten, then these species have the potential to become established.

Today, conservation efforts focus mainly on species whose introductions cause ecological or economic harm. Scientists intervene unabashedly, trying to thwart these species' establishment by creating environments in which they are unfit. Strategies include poisoning invasive species with chemical herbicides and pesticides, often trading temporary reductions in invasive species abundance for deteriorating water and soil quality, or introducing species that might eat or outcompete the more destructive species. Scientists have even tried to remove invasive species by hand.

Some of these efforts have been impressively successful. In 1993, scientists released ladybird beetles on the small island of St. Helena in the south Atlantic Ocean, where South American scale insects, which had been introduced two years earlier, were destroying native gumwood trees. The ladybird beetles were efficient predators of the scale insects. No scale outbreaks have been reported on the island since 1995, and native gumwoods are thriving. A concerted hunting program led by the US National Park Service eradicated feral pigs from Santa Cruz Island, a small island off the coast of California, in just over a year beginning in 2005. With pigs absent, the island's native plants and animals, eight of which are listed as endangered, have begun to rebound.

While manual removal has the least potential of our existing strategies to cause long-term ecological harm, hard-won reductions in invasive species' populations can be short-lived. From 2017 to 2019, hunters were paid to spend their nights roaming the Florida Everglades in search of invasive Burmese pythons. Invasive pythons have been a screaming evolutionary success in the Everglades. Their camouflage is ideal, and they eat everything from small mammals and birds to white-tailed deer and even alligators. Unsurprisingly, Burmese pythons have had a disastrous impact on local wildlife, and in particular on birds. Over the two-year period of their campaign, the hunters captured more than 2,000 Burmese pythons in the Everglades, some longer than 5 meters (16 ft). Their effort, however, barely made a dent in the invasive python population. We need another solution.

In early 2019, the not-for-profit organization Island Conservation reported that they had successfully removed all rats from the Marquesan island of Te Uaua. This marked the 64th island from which Island Conservation had successfully eradicated invasive rats. The ecological recovery in each case has been astounding. With rats no longer eating their eggs, native seabird populations rebound. Native plants also return, as seeds and young plants go uneaten. People benefit from increased crop yield and the elimination of rodent-borne disease. Island Conservation's approach to rat removal, however, is controversial. Their approach relies on rodenticides, blanketing islands with rat poison pellets dropped from helicopters and drones. Rodenticides are effective but can have unpleasant side effects. Rodenticides can't target species individually and may harm birds that eat poisoned rodents. And any chemical or toxin has the potential to contaminate soil and water. For now, the local communities believe that the potential dangers of rodenticides are worth the rewards. The Island Conservation team agrees but is also looking toward the future, where it sees a new, safer solution in synthetic biology. It's joined forces with an international team of scientists and nonprofit organizations to form the Genetic Biocontrol of Invasive Rodents program. The program's goal is to engineer rats that cannot reproduce by inserting a mutation into their DNA that makes them sterile.

As an approach to invasive species management, synthetic biology promises to be more efficient than manual removal and both more humane and safer for the environment than poisoning. It also, however, pushes us deeper into the role of masters of other species' evolutionary future. This is, of course, a role into which we have already been cast. The question now is how far we will go. Shall we allow ourselves to alter a species' DNA directly in order to save it or another species from becoming extinct? How different is this approach, compared to those that we already use?

As we consider these questions and the options now before us, volunteers throughout the swampy Southeast wade every year into local ponds, lakes, and rivers to take part in planned cleanups. Waist-deep in murky stale water, they grab handfuls of invasive water hyacinth that they stuff into large plastic buckets, clearing room for kayakers, swimmers, and fish. And every year the water hyacinth grows back.

— PART II —
The Way It Could Be

— CHAPTER 6 —

Polled

O N AN EARLY AUTUMN AFTERNOON IN 2019, I CLIMBED into the passenger side of Alison Van Eenennaam's Honda CRV in downtown Davis, California, joining a small crew from her lab for the short ride to the University of California, Davis, beef barn. The license plate on the SUV read "BIOBEEF" and a green plastic crocodile, nostalgic of Alison's Australian heritage, stood guard along the dash. "This is the beefmobile," Josie Trott, Alison's lab manager and right-hand woman, told me as she settled in behind the wheel. "Alison told us not to crash it."

I was in Davis to deliver the Genomics Center's Halloween lecture the next morning, which I'd agreed to in part because it meant an opportunity to meet Alison. Alison is one of the leading scientists using biotechnology to advance animal agriculture, and also an adept communicator and ambassador of this research. I wanted to pick her brain not only about the details of her work but also about her outreach to the general public, which, I think it is fair to say, is not entirely on board with the use of biotechnology in agriculture. And of course, after the events of the last few months, I wanted to meet her prized cow, Princess. Unfortunately for me, Alison's schedule was incompatible with this plan, as she was spending Halloween in Australia at an animal breeding and genetics conference. In her place, though, Josie offered to introduce me to some of the others in their group and, perhaps, take me to meet some cattle.

Our first stop was lunch. As we waited for our food, Josie and the two others who joined us, Joey Owen and Tom Bishop, told me about their ongoing projects. Joey and Tom were exploring different ways to use biotechnology to give farmers more control over whether male or female calves were produced. Dairy farmers, for example, might prefer to produce mostly female calves, since males don't grow up to produce milk. A genetic approach that limited the production of males would save them the expense of selling males for food or destroying male calves after birth.

As I listened to Joey and Tom and thought about Princess, it was clear to me that new biotechnologies—molecular tools that turn genes off, change letters in the DNA code, take a gene from one species and insert it into the genome of another—have brought us to the precipice of another major transformation in our relationships with the plants and animals around us. Less clear was how far people working in synthetic biology believe we are from that cliff's edge. Today's motivations for using new technologies are the same as those that drove our ancestors: to improve human lives, the lives of our domesticated animals, and the environments in which we all live. But the new technologies feel different, less natural perhaps. Worse, their use is surrounded by an amorphous uneasiness that is fed and compounded by well-funded global campaigns, rife with disinformation, that intentionally cloud understanding of what these biotechnologies can and cannot do. The consequences of these campaigns cannot be understated; today's environment is so polarized that simply discussing a project that uses these tools can incite mistrust, anger, even violence.

Alison's research is an exemplar of the difficulties of working in this new space. Her goal is to improve animal welfare—to reduce suffering—while at the same time improving the economics of cattle farming. And despite spending her career campaigning for these technologies, Alison is despondent about the future.

As Joey and Tom explained their experiments to me, their excitement, too, was tempered. I sensed that they were doing their best to stay upbeat, to press on with their work despite the setbacks the team had faced over the previous few months. But after dealing with the public reaction to what happened with Princess, they were struggling to imagine a future where their work could have the impact that they wanted it to have.

We finished lunch and headed back to the SUV. Next stop: the beef barn, and (I hoped) Princess. On the road, we chatted about careers in science and biotechnology and about where all of this might be going. Joey, about to finish his PhD, was torn between a future in academic research and one in industry. This is not an easy choice. While working in a university setting might allow more creative independence, funding for academic research like that in Alison's lab is nearly nonexistent, due largely to the unclear regulatory path for biotechnologies in agriculture. Without a regulatory framework, scientists like Alison, Josie, Joey, and Tom might spend years working on a project only to learn at the last minute about a change of heart, a reinterpretation of a rule, a new hurdle to climb. Public institutions like UC Davis simply don't have pockets deep enough to pay for the endless series of experiments mandated by today's shape-shifting regulatory environment. For Joey to keep doing this kind of work, his only choice might be to take a job in industry.

Our small group fell quiet as we pulled up to the barn. I was thinking about Princess and her struggles within the existing regulatory environment. Princess was (at least to my mind) a normal dairy cow. But she was also a product of an experiment that I knew from media reports had not gone entirely according to plan. I knew her male siblings had been incinerated and that she alone was spared, for now. Would it be obvious when I saw her that she had been genetically engineered?

We climbed out of the SUV and headed through the main barn structure to the far door, then back outside into the heart of the farm. There were cattle everywhere. In the distance, a herd was grazing the dregs of last season's grass. Nearby, a maze of dozens of pens separated cattle by age and breed and, presumably, experiment. We passed a pen with two cows and their young calves, one of which was working determinedly but with little success to get milk out of either cow. We turned a corner and headed past a pen of maybe 20 heifers chewing their cud and staring blankly into the distance, and then another of the same number of equally placid-looking bull calves. Finally, we got to Princess, in the third and last pen of that row. Hers was the same size as the other pens but she shared her space with only one other animal—a robust bull. Both stared intently at us as we walked up and stopped in front of them.

"That's her husband," joked Josie, motioning to the bull. And then she clarified: "His job is to get her pregnant." Josie reached into the trough to separate clumps of alfalfa from oat grass hay, offering the alfalfa to Princess as a treat.

I looked confused. "I thought the experiment was over?" I asked delicately.

"We have to test her milk," Josie explained. "For the FDA."

"Test her milk for what?" I wondered aloud.

Josie turned to me, the look on her face a mixture of frustration and bewilderment. "Exactly," she said, as Tom tutted and turned away.

GMOS ARE BAD! WHAT'S A GMO?

It is not an understatement that genetic engineering is a contentious topic in agriculture. Some people are decidedly against using the tools of synthetic biology to modify food crops and animals, citing the "unnaturalness" of the process and its consequent unpredictable riskiness. Others view genetic engineering as nothing more than a quick and precise means to manipulate species exactly as we've been manipulating them since the origin of farming. The truth is somewhere in the middle.

The intended impact of genetic engineering is the same as that of traditional breeding selection: to create tastier or more useful plants and animals. The process, however, is different. In the traditional approach, we breed two individuals and hope that some offspring show the intended trait. In genetic engineering, an organism's DNA is edited directly, guaranteeing the trait's appearance in the next generation. This makes genetic engineering quicker than selective breeding, sometimes shortening the process by decades. And given the growing number of people alive on the planet, all of whom need to eat, a faster and more efficient approach to refining our crops and domesticated animals is welcome.

While the end products of genetic engineering are often identical to what might be expected after generations of selective breeding, they don't have to be. Genetic engineering can create what are called transgenic plants and animals, organisms that combine traits from different species. Transgenic organisms sound like science fiction but are not. Today, transgenic crops express bacterial genes that make the plants insecticidal. Transgenic nanny goats and cows express human genes that alter the composition of their milk, enhancing its antimicrobial properties or making it drinkable by people with allergies to non-human milk. Transgenic Hawaiian papayas express viral genes that make them immune to papaya ringspot virus. And these are just a few examples of real transgenic organisms.

During the early days of synthetic biology, most genetically engineered organisms—even those for which the intended end product was something that could be generated using traditional breeding—were at least a little bit transgenic. This is because it was common to integrate fragments of bacterial DNA as a way to check that the intended edits occurred (we'll talk more about this later). This led to the broad-brush characterization of genetically engineered foods as *Frankenfoods*, an unshakable epithet prompted by a 1992 letter to the *New York Times* in which Boston College English professor Paul Lewis objected to the idea of genetically modified tomatoes.

More recent approaches to genetic engineering leave no trace of the editing process in the genome. As a consequence, new genetically engineered organisms tend to be cisgenic rather than transgenic, meaning that they contain no DNA from other organisms. To distinguish them, these products are often referred to as "gene-edited" rather than "genetically engineered." Cisgenic genetically engineered gene-edited organisms also follow simpler paths to market than transgenic genetically engineered organisms, and many companies have shifted from creating new combinations of organisms to (transgenics) exaggerating or deleting existing traits.

While many gene-edited organisms are essentially identical to the products of traditional breeding, people continue to associate all genetically engineered organisms, sometimes called genetically modified organisms or GMOs, with transgenics, and often really disgusting-sounding transgenics—like Lewis's envisaged GMO tomatoes augmented with fish DNA, which didn't exist then and don't exist today. This knee-jerk yuck factor has allowed confusion and disinformation about GMOs to proliferate. Food producers and marketers hoping to profit from customers' discomfort adorn their products with reassuringly bright green labels that declare them "non-GMO," even when there is no genetically engineered product to differentiate. On a trip to the grocery store a shopper will encounter non-GMO labels on citrus fruits, tomatoes, kidney beans, olives, and a long list of other products for which there are currently no GMO options to purchase. They may be surprised to also find non-GMO labels on salt, which is a mineral and has no DNA to modify. What exactly does the bright green non-GMO label mean? Not very much, it turns out.

Defining a GMO is tricky. I could argue, and some people do, that everything we eat is genetically modified. This is true in a sense, given the thousands of years of breeding selection that created our domesticated plants and animals. But this is also not the definition intended by the term GMO, and misconstruing it as such dismisses the genuine differences between genetic engineering tools and traditional breeding.

A slightly narrower definition of GMO includes only organisms developed by people using means other than traditional breeding. This definition is, however, still too broad. Foods like honeycrisp apples, navel oranges, seedless watermelons, and hazelnuts are not products of traditional breeding, but neither are they genetically modified. Instead, these products were created by grafting—splicing together body parts of different species or lineages of plants. Grafting is crucial to production of some of our favorite non-GMO-labeled foods. Wine grapes, for example, are under constant attack by phylloxera, a disease caused by aphids. People introduced the aphids to Europe accidentally from the Americas during the nineteenth century, nearly shuttering the European wine industry as the aphids quickly destroyed the vines. When European vines were grafted onto roots of phylloxera-resistant American grape varieties, however, the European vines survived, and the wine continued to be delicious. Today, nearly all wine grapes grown around the world are grafted onto American rootstocks, but few people would argue that these wines should be labeled as GMOs.

Because grafting doesn't affect the DNA within individual plant cells, grafted organisms can be excluded as GMOs by narrowing the definition to focus explicitly on organisms with modified DNA. With this in mind, the European Union defines GMOs as organisms with DNA "that has been altered in a way that does not occur naturally by mating or natural recombination." Intriguingly, however, this definition excludes the many varieties of fruits, vegetables, and grains that have been produced over the last century by mutation breeding—a strategy for generating new plant varieties by exposing seedlings deliberately to mutation-inducing radiation or chemicals.

Mutation breeding causes lots of changes to the DNA sequence randomly across the genome and, in doing so, changes the phenotype of the plant. Brown rice, the popular disease-resistant Renan wheat, and Ruby Red grapefruits are all products of mutation breeding, but, as indicated by the bright green label on the bottle of Ruby Red grapefruit juice in my fridge, not considered to be GMOs. Why not? The European Union argues that while, yes, a lot of mutations occur at once during mutation breeding, given enough time and exposure to natural mutagens (like UV light), any of the useful mutations would probably emerge on their own. Because these new varieties could occur naturally, they are therefore not covered by the European Union's definition of GMOs.

To be clear, I am not arguing that products of mutation breeding should be considered GMOs. Nor do I think these products should be subjected to additional regulation compared to traditionally bred varieties. Mutations are not inherently dangerous. Each time a cell divides, a new copy of that cell's genome is made, and that copy contains a few mistakes. Every human child, for example, has 40 or so new mutations in their genome that neither of their parents had, most of which have no impact on the child. Instead of arguing for increased regulation of products of mutation breeding, I bring these products up to highlight the hypocrisy of ignoring the thousands of uncharacterized and random genetic alterations that emerge through mutation breeding while excluding products from the market that contain few, specific, and deliberately induced mutations on the grounds that unintended consequences of these mutations might turn out to be dangerous.

While the European Union's definition of a GMO focuses on the process of engineering an organism, the United States has elected to regulate the end product. This doesn't mean, however, that all genetically engineered organisms are treated equally. In the United States, three agencies regulate genetically engineered organisms in what is known as the Coordinated Framework. The US Department of Agriculture, or USDA, regulates plants; the Federal Drug Administration, or FDA, regulates animals and animal feed; and the Environmental Protection Agency, or EPA, regulates pesticides and microorganisms. Each agency has adopted a slightly different regulatory approach. The FDA, for example, regulates genetically engineered animals and animal food as drugs, requiring the same assessments of safety and efficacy as required of a new cancer drug. The USDA, alternatively, has chosen not to regulate genetically engineered plants if the final product is indistinguishable from a traditionally bred plant. While the USDA decision creates a less restrictive environment for developing gene-edited plants in the United States than exists elsewhere, the lack of a coordinated global framework for regulating these products is untenable in the long term. What will happen when a plant developed in the United States using gene editing is released onto a European farm? Does it suddenly become a GMO? And, since the final product is unrecognizable as a GMO, how would anyone be able to tell? More importantly, does it really matter?

HORNLESS HOLSTEINS

Buri, Princess's father, was born in 2015 on a farm in Minnesota. He was one of several bull calves born that spring using a process called somatic cell nuclear transfer or, more commonly, cloning. Cloning involves creating an entire organism starting not from the cell that forms when a sperm fertilizes an egg, but from another cell—a somatic cell—taken from some other tissue in the body. At a super basic level, cloning happens like this: An unfertilized egg cell is harvested and its nucleus—where the DNA is—is removed and replaced with the somatic cell's nucleus. Then, in a step called reprogramming, proteins in the egg cell trick the somatic cell's genome into forgetting how to be whatever type of cell it was (a skin cell or mammary cell perhaps) and reverting to the type of cell that forms when a sperm fertilizes the egg—the type of cell that can divide and differentiate into all the different cell types that make up a whole organism. Cloning has been commonplace in cattle agriculture and, by 2015, the birth of healthy cloned calves was welcome but not particularly earth-shattering news. Buri, however, was not only a clone. He was a gene-edited clone.

Buri's genome was edited by scientists at a biotech company called Recombinetics. Their goal was to delete, in an early stage embryo, a small string of DNA letters on cattle chromosome 1 and replace it with a different, slightly longer string of DNA letters. Individuals' genomes often have slightly different versions of the same stretch of DNA, and these versions are called alleles. If Recombinetics was successful in swapping one allele for the other, the resulting animal would not grow horns.

Hornlessness, which is also known as "polled," has been observed in domestic cattle for millennia. The oldest evidence of polled cattle is from ancient Egyptian art depicting children milking cows without horns—a sign that hornlessness might also have been associated with docility. Dozens of European archaeological assemblages dating to within the last 4,000 years include hornless cattle skulls, suggesting that farmers across cultures were choosing hornless cattle over their horned brothers and sisters. The allele that Recombinetics was working with is estimated to have evolved just over 1,000 years ago, and it is one of several polled mutations found in today's cattle breeds.

It's easy to imagine why herders and farmers throughout history preferred hornless cattle. Hornless animals are easier to herd, move, and milk. Horns are sharp, and being poked with them can hurt both other cattle and any person who happens to get in the way. Hornless cattle can live in higher densities, and a cattle farmer whose net worth is determined by the number of heads of cattle on his land can accommodate more heads when those heads lack horns. Hornless cattle are so highly valued today that farmers often choose (or are compelled by legislation) to remove horns surgically.

In the United States, around 15 million calves are surgically dehorned each year. Dehorning is an unpleasant, expensive, and painful process that, unsurprisingly, raises significant concerns related to farm animal welfare. It is also precisely the process that Recombinetics was aiming to eliminate, or at least reduce, when it engineered Buri's genome. By swapping the hornless (polled) allele from Angus—an elite beef breed that happens to have evolved hornlessness—into the genome of a Holstein—the black-and-white breed that dominates the dairy industry—Recombinetics aimed to create a hornless Holstein bull that could then be bred with Holstein cows, increasing the frequency of hornlessness in this agriculturally important breed.

But wait. Hornlessness already exists in Angus. In fact, many cattle breeds have some proportion of naturally hornless individuals, including some dairy cattle. And any of these could be bred with Holstein. Why not just do this the normal way?

Because that would be an environmental and financial disaster.

Traditional breeding or artificial insemination could be used to pass the polled allele from Angus to Holstein. If a Holstein cow was inseminated with sperm from a polled Angus bull, the calf would inherit the polled allele from her father and, because only one copy is necessary to have the intended effect, would be hornless. The trouble, of course, is that the polled allele is not the only DNA that she would inherit from her father. Indeed, a full half of her genome would come from him, meaning that one copy of every gene would be a version optimized for beef. This would be terrible for the dairy farmer. Today, elite Holstein cows produce 25 percent more milk than they did ten years ago while requiring less food, water, and space. Because more of their feed is going to milk production, they also produce less manure and less methane. When a Holstein cow breeds with an Angus bull, all this optimization is lost. The calf's genome is a random blend of Holstein and Angus alleles, and she would be neither an efficient dairy cow nor an elite beef animal. The high-value, dairy-specific traits could be restored by selectively breeding hornless but less optimized Holsteins with elite Holsteins over several generations, but this would take decades during which the dairy farmer would experience significant economic loss.

Rather than blend two genomes randomly and hope for the best, gene editing allows precise, targeted, selective breeding. We know exactly what genetic change we want to make to induce the desired phenotype—hornlessness—and we can make that change with pinpoint precision. With gene editing, we can move the naturally occurring hornless phenotype from Angus to Holstein in a single generation, improving animal welfare without disrupting the traits that make Holstein cows fantastic milk producers. The gene-edited hornless Holstein is not transgenic—the trait evolved naturally in cattle. And because the hornless allele has been in our cattle for hundreds of generations, we know exactly what phenotype to expect: a healthy, fertile, hornless animal whose meat and milk are as safe to consume as they have been for thousands of years.

All of this sounds great, right? Anyone hearing of hornless Holsteins as their introduction to this family of new biotechnologies might even wonder what all the fuss is about. But this is not where the story of genetic engineering technologies, or the opposition to them, begins. For that we need to go back in time almost 50 years.

"NOW WE CAN COMBINE ANY DNA"

Herbert Boyer let the cat out of the bag at a scientific conference in 1973. It was probably an accident; Boyer had been invited to talk about his lab's discovery of *Eco*RI, one of a family of newly identified molecules called "restriction enzymes" that were allowing scientists to study DNA in unprecedented detail. *Eco*RI was core to the story

Boyer told, but it was the other details, those that he was not sup-posed to divulge, that caught the audience's attention and set into motion the chain of events from which we are still recovering.

Boyer was a biochemist from the University of California, San Francisco, whose lab was among the first to isolate and describe restriction enzymes. Restriction enzymes can be thought of as mo-lecular scissors that evolved to find and cut specific sequences of DNA. As was common in the 1970s, Boyer shared *EcoRI* widely after its discovery so that other labs could try it in their own re-search. Of relevance to this story is that he sent *EcoRI* to Paul Berg, a biochemist at nearby Stanford University.

Berg's lab was developing tools to discover gene functions. One way to do this is to add a gene into a cell's genome and measure whether that cell changes in some way, maybe by producing more of a protein or by growing at a different rate because of the addition of that gene. Berg could culture cells (make them grow into colonies) in dishes in the lab, but he needed a way to move the genes that he wanted to study into those cells' genomes. This is where *EcoRI* and its DNA-chopping ability come in. Berg hypothesized that *EcoRI* would cut the genome open so that other DNA could be inserted. Then he could use another newly discovered molecule—a ligase—to solder the DNA back together.

Berg planned to splice together the genomes of two viruses: SV40, a small and well-studied virus that infects monkeys, and lambda virus, which infects bacteria. The choice of lambda was key. While viruses like SV40 make copies of themselves by hijacking components of the host's DNA replication machinery, lambda virus makes more of itself by inserting its genome into the host's DNA directly. If Berg were successful in splicing the two viruses together, the lambda virus would insert the combined virus genome into that of the host cell. If the experiment succeeded, he would have devel-oped a new system to insert DNA into genomes, and the perfect system for studying gene function.

In 1972, Berg's lab cut open the circular genomes of SV40 and lambda virus and spliced the two virus genomes together. This created the world's first "recombinant DNA"—a genome engineered to have combined (or, in the parlance of genetics, recombined) DNA from more than one organism. They intended to insert their recombinant DNA into the bacterium *Escherichia coli*, as lambda virus naturally infects *E. coli*. Before the experiment was scheduled to take place, however, Janet Mertz, a graduate student and key member of Berg's team, disclosed the plan to scientists at the Cold Spring Harbor Lab, where she was taking a course. Their reactions were harsh. *E. coli*, they pointed out, grew readily in the human gut and SV40 was known to cause cancer in small mammals. In doing this experiment, they wondered whether Berg's team would be putting itself, and possibly the rest of the world, at unnecessary risk. Mertz communicated these concerns to Berg, who surveyed other researchers and found that many felt similarly. He stopped the experiments. This was important work, but safety was paramount.

While Mertz, Berg, and others were splicing viruses together, Stanley Cohen, another Stanford scientist to whom Boyer had sent *Eco*RI, was exploring whether *Eco*RI could splice together bacterial plasmids—small circular DNA molecules that bacteria exchange with one another as a way of swapping genes. To Cohen's satisfaction, *Eco*RI could indeed cut some bacterial plasmids. Taking advantage of this discovery, Boyer and Cohen recombined the DNA from two bacterial plasmids and took the next step, injecting the recombined plasmid into *E. coli* cells. They'd selected plasmids that made *E. coli* resistant to antibiotics, in fact each plasmid contained resistance genes to a different antibiotic. This meant that they could test whether their experiment was a success by treating the (hopefully recombined) *E. coli* with both antibiotics. If the colonies survived, then they would know that both plasmids had gotten into the bacterial genome.

When Boyer and Cohen treated their bacteria with antibiotics, the colonies survived. Their experiment was a success. Although they did not use today's much maligned term, they had created the first self-replicating genetically modified organism.

It was this experiment—the successful splicing together of plasmids and insertion of this recombined plasmid into *E. coli*—that Boyer revealed accidentally at the 1973 conference. From the back of the room, someone is said to have shouted, "Now we can combine any DNA." But, as with Mertz's revelation of the virus-splicing experiments at Cold Spring Harbor, excitement was not the dominant reaction. The scientists at the meeting were nervous. DNA could be spliced together, and that was certainly cool. But these were only the first experiments, and scientists had already engineered potentially cancer-causing viruses and made bacteria resistant to multiple antibiotics. This was obviously powerful technology, but how powerful? The scientists wanted to learn more, but they also wanted to keep everyone safe.

By the end of the conference, the attendees had written and sent a letter to the National Academy of Sciences and the Institute of Medicine, asking them to establish a committee to consider the hazards of recombinant DNA research. The letter underscored the potential of recombinant DNA experiments to advance science and improve human health but also raised concerns over the still-unknown outcomes of recombining DNA in a lab. The scientists wanted a better understanding of what controls and containment protocols should be in place to protect people working in the lab and to protect the public at large. They wanted to be proactive rather than reactive.

The next steps were taken immediately. A committee was formed, a temporary moratorium was declared on research creating recombinant organisms, and an international meeting was planned to decide the future of recombinant DNA research. While these measures were intended to assuage public concerns, they—unfortunately—had the opposite effect. Sensing that scientists feared the worst, protests began against recombinant DNA technology even before the technology could be evaluated. Jeremy Rifkin, who takes credit for starting the anti-GMO movement, raised money by frightening the public into believing that people were going to be cloned (recombinant DNA technology is not cloning). Concerned members of the public lobbied their government representatives to stop the research. By the time the meeting took place, a clear line had already formed between those who wanted recombinant DNA research to succeed and those who wanted to ban it from happening at all.

The meeting to decide the future of recombinant DNA technology was held in February 1975 at the Asilomar Conference grounds in Pacific Grove, California. Attendees included scientists, ethicists, and legal scholars. Most supported allowing recombinant DNA research to continue, but not without hesitation. They worried about what might happen if plant or animal genes were inserted into a bacterial genome. Could the new genes cause bacteria to make something toxic to plants or animals? If an animal ate recombinant bacteria, could the new genes jump into that animal's genome, potentially harming the new host? In the end, the assembled group agreed that the research had tremendous potential and should continue. They insisted, however, on strict safeguards and containment protocols for potential biohazards. Participants left the meeting feeling as if they'd paved a way forward for safe recombinant DNA research.

The conclusions of the Asilomar Conference were reported in both the scientific and popular press. The scientists who participated were pleased with the consensus and believed the public would be as well. Not so. As before, anti-biotechnology activists exploited the risk-averse outcome of the meeting to further undermine public trust. Rumors spread that recombinant DNA technology would soon generate superbugs or be used to create superhumans. The division between supporters and opponents deepened.

After Asilomar, recombinant DNA research recommenced under intense scrutiny. In Cambridge, Massachusetts, local politicians insisted that researchers use containment facilities designed for airborne infectious diseases, despite the fact that *E. coli* does not spread via the air and, even if it had escaped, the strain used in the lab would not be able to live in the human gut. Scientists had no choice but to go along with these restrictions, even though doing so reinforced public perceptions of the research as high risk. Despite these challenges, however, the practical power of recombinant DNA technology was indisputable. Bacteria could be coaxed to do new things, to express genes that humans engineered them to express. Scientists could use recombinant DNA technology to learn the functions of genes, accelerating the genome's decoding. And by turning bacteria into living protein factories, the technology could reduce our reliance on animals for biological products.

Within three years of the Asilomar meeting, a biotech start-up called Genentech, founded by Boyer, discovered how to engineer bacteria to express human insulin, the protein that regulates the amount of glucose in blood. People with type 1 diabetes cannot produce insulin on their own and must inject it in order to stay alive. Before recombinant insulin became available, insulin was harvested from pig and cattle pancreases, for which more than 50 million animals were slaughtered every year. Eli Lilly, the drug company that sold most of the insulin on the market, immediately saw the value in recombinant insulin. Eli Lilly bought the technology from Genentech and scaled production, quickly outpacing production from animals. Clinical trials of recombinant insulin began in 1980 and were an astounding success. The insulin worked as intended, and some diabetics who'd reacted poorly to animal insulin improved when they switched to the human version produced by recombinant organisms. The era of synthetic biology had begun.

RECOMBINANT PLANTS

Although the medical industry was first to embrace recombinant DNA technology for its commercialization potential, the agricultural industry was not far behind. Before this could happen, though, scientists needed to discover a way to recombine DNA into plants. Fortunately, a family of bacteria evolved to do just that.

Agrobacteria are soil-dwelling bacteria that infect plants by invading wounds on their roots, stems, and leaves. Once they get into a plant cell, they insert a bit of their own DNA—a plasmid—into the plant's genome, just as lambda viruses and bacterial plasmids insert into bacterial genomes. The infected plant then expresses the genes in the newly inserted *Agrobacterium* plasmid as if those genes were the plant's own. But they're not. They're invaders. The *Agrobacterium* genes cause the plant to produce a tumor-like gall within which the bacteria will live and multiply. They also cause the plant to make hormones that disrupt that plant's ability to fight the disease and molecules called "opines" that the bacteria use to make more bacteria. This is a pretty neat trick, as long as you're not rooting for the plant. It's also precisely the trick to manipulate to engineer recombinant plants.

Scientists now know which parts of the *Agrobacterium* plasmid's DNA are necessary for it to integrate into the plant genome. They also know which bits cause disease and, thanks to recombinant DNA technology, can cut these bits out (because they don't want to make the plant sick) and insert other DNA in their place. Then they can use *Agrobacterium*'s naturally evolved mechanism of infecting wounded plants to insert the modified plasmid into the plant's genome.

In 1983, in a single session of a biochemistry conference known as the Miami Winter Symposium, three separate research groups that had been competing with one another for years announced in back-to-back presentations that they had successfully engineered plant genomes using *Agrobacterium*. All three had removed the disease-causing bits from an *Agrobacterium* plasmid and inserted a gene that would make plants resistant to antibiotics. The antibiotic resistance gene acted as a "marker" and allowed them to learn which, if any, plant cells had been infected and altered. Over the next year, each of these labs published manuscripts describing their approaches to engineer plant cells.

The years following the 1983 Miami Winter Symposium saw huge investment in developing recombinant DNA technologies for agriculture. Academic and commercial labs worked to discover which plant genes cause what traits (What gene causes potatoes to turn brown?), to develop genetic tweaks to alter gene functions (How can we turn off the gene that causes potatoes to turn brown?), and to improve the efficiency of *Agrobacterium*-mediated DNA transfer (How can we get our tweaked gene into the potato genome?). The so-called gene gun, developed in 1987, was key to accelerating innovation. Before the gene gun, scientists relied on the natural and haphazard infectivity of *Agrobacterium* to get the modified plasmid into plant cells, but too few plant cells were being infected. The gene gun shoots particles coated with plasmid DNA into plant tissues directly, improving the rate of plasmid integration. In fact, when using the gene gun, modified DNA often integrates into each plant genome several times.

Soon, crops engineered using the *Agrobacterium* system were being grown on farms rather than in greenhouses. The first of these were engineered for traits that benefited farmers. In 1986, tobacco plants engineered to be resistant to herbicides were tested simultaneously on farms in France and in the United States. Farmers growing these engineered plants were able to trade less effective and environmentally persistent herbicides for more potent, rapidly degrading herbicides. A year later, the first Bt crops, which are plants engineered to express a gene from the bacterium *Bacillus thuringiensis*, were planted. Bt produces a protein that is toxic to insects, allowing farmers to reduce their use of insecticides. Soon after, China became the first country to commercialize a genetically engineered crop, with the approval for sale of tobacco engineered to be immune to tobacco mosaic virus, which causes the leaves of infected plants to become wrinkled and discolored, stunting the plant's growth and reducing profit.

Because the early plant engineering experiments targeted improvements in crop production rather than quality, the public—the eventual consumers of these products—were largely left out of discussions about the science. It wasn't obvious to the average person why genetic engineering was useful or what they might gain personally from further development of these technologies. No one communicated successfully, for example, that years of research confirmed that Bt was toxic only to some insects and not toxic to humans or other mammals. Few people knew about the controls in place to limit the ecological impact of planting genetically engineered seeds. Meanwhile, the misinformation campaigns against the technologies intensified. False facts maligning the new crops spread with limited effort to correct the public record. What was needed, desperately, was a genetically engineered product that targeted the consumer rather than the producer: a high-profile excuse to engage the public about the science and the safety of genetically engineered plants. The industry needed a story that could out-PR the anti-GMO activists. That story would be about a tomato: a superbly delicious, plump, firm tomato that could be found on grocers' shelves even in deepest winter.

A TASTIER TOMATO

I love tomatoes. Tiny, sweet ones in particular. But also big beefy ones and oddly shaped heirloom ones and unusual speckled green ones. But every now and then I get a tomato that is just, well, meh. Often my disappointing tomato looks delicious—bright red, taut flawless skin, perfectly plump. But when I bite into it, the fruit is soft and mealy, or too watery, or just tasteless. Fortunately, tomato disappointment is rare today. But in the 1980s and 1990s, nearly all tomatoes on grocery store shelves were disappointing, particularly during the off-season. But everyone wanted non-disappointing tomatoes all year round, and this is why tomatoes were the ideal candidate for genetically engineered improvement.

The problem to overcome was the tomato's notoriously short shelf life. The deliciousness of a freshly picked ripe tomato lasts only a few days before the flesh starts to go soft and the fruit begins to rot. A solution is to grow masses of tomatoes in warmer parts of the world and pick them while green and hard as a rock, as unripe green tomatoes can be stacked and shipped long distances. Then, just before transfer from storage to sale, expose the fruits to ethylene gas, which mimics natural ripening cues and causes the fruit to turn appetizingly red and begin to soften. Looks, however, are deceiving. Ripening with gas rather than on the vine results in tomatoes that taste disappointingly green.

In August 1988, Calgene, a small biotech company from Davis, California, announced that it had solved the problem of the tasteless tomato using the power of genetic engineering. Scientists at Calgene and elsewhere had seen in their greenhouses that a protein called polygalacturonase, or PG for short, increased in concentration as tomatoes ripened. They also observed that mutant tomatoes that ripened without softening had little PG in their fruit. These observations led to the hypothesis that PG is what caused softening, so Calgene set out to invent a way to stop expression of PG during ripening. Their goal was to create a tomato that turned red but didn't rot.

Calgene scientists controlled PG expression by inserting an extra copy of the PG gene into the tomato genome that was backward and flipped around compared to the original copy. Called "antisense," this back-to-front extra copy of the PG gene blocked the original copy from making PG. The resulting tomatoes did not accumulate PG during ripening and, most importantly, stayed firm for weeks longer than traditional tomatoes did after being picked. Its new tomato, Calgene conjectured, could ripen on the vine and be shipped long distances. Goodbye, gassed green tomatoes!

When Calgene told the world about its long-lasting tomato, which it would later name the Flavr Savr tomato, the road to getting that tomato into sauces and salads was long. Calgene was a small company, and it had no precedent to follow to get a genetically engineered food to market. There were rival groups to compete with, at least one of which had also developed antisense PG technology. There were board members to satisfy with positive financial projections and lawyers to appease with carefully curated lab notebooks. There were skills to learn related to growing and shipping tomatoes. And there was a path to pave toward commercialization of the world's first genetically engineered food intended explicitly for human consumption.

Calgene was of course aware of the anti-GMO movement and the widespread mistrust of genetic engineering that this movement sowed. However, its tomato differed from other GMOs in ways that it believed the public would appreciate. The tomato was engineered neither to express bacterial toxins or viruses nor to be resistant to herbicides. And the problem that the tomato solved—the common experience of being disappointed with gassed green tomatoes—was one to which people could relate. Perhaps Calgene's tomato would be the product to break through the anti-GMO wall.

Calgene believed that the path to public acceptance of the Flavr Savr tomato was regulation, or more specifically deregulation: declaration by the agencies charged with protecting the public that the product was not only safe but so similar to traditional varieties that additional regulation was unnecessary. Given the aggressiveness of the anti-GMO movement, the company also knew that people would be suspicious of its product and its intentions. Calgene decided to address this head-on by making all its interactions with the agencies—their petitions, their data, descriptions of their experiments—open and accessible. Calgene wanted to tread carefully. It wasn't only a tomato that was on the line but the future of an entire industry.

Calgene began its long walk toward deregulation by working to prove the safety of what it thought would be the most controversial aspect of the tomato: antibiotic resistance genes. As did most genetic engineers at the time, Calgene included antibiotic resistance marker genes in its *Agrobacterium* plasmid to provide a quick test that their experiment was successful, which in this case meant that the antisense PG gene had also been inserted into the tomato genome. But this meant that both of these genes, the antisense PG gene and the antibiotic resistance gene, would be expressed in every tomato cell. To convince the FDA that consuming tomatoes with antibiotic resistance genes in their DNA posed no heightened risk compared to eating tomatoes without them, Calgene would have to imagine all the ways that eating antibiotic resistance genes might harm someone and then test explicitly whether any of those ideas panned out.

What are the risks associated with eating antibiotic resistance genes? A first hypothesis is that if a person or animal were to eat antibiotic resistance genes they might become resistant to antibiotics. The route to this scary outcome is not via integration into our own DNA (all the food we eat contains DNA, but we don't worry that we might become part cattle by eating a burger), but whether the DNA from something we eat survives digestion and integrates into the genomes of the bacteria in our guts. Might this be possible? Or would the DNA instead degrade away to nothing as part of the digestion process? Calgene would have to find out.

To measure how quickly the DNA that we eat degrades, Belinda Martineau, a member of Calgene's science team, exposed DNA to simulated digestive fluids. After ten minutes in simulated stomach juice and ten minutes in simulated intestinal juice—a much shorter transit time than our food normally takes through our digestive systems—she measured the DNA that remained. No DNA fragments survived that were as long as an antibiotic resistance gene. Because a gene needs to be intact to function, her results meant that the chance that an antibiotic resistance gene could survive digestion and become incorporated into the genome of a microbe living in our intestines was very, very, very low. But how low? To approximate a number, Martineau measured the length distributions of surviving fragments of DNA and guessed, conservatively, that for every 1,000 people who eat a Flavr Savr tomato, one intact antibiotic resistance gene would make it to their gut intact, which would make it possible for a microbe to incorporate that gene into their own genomes. Since all of us have billions of gut microbes, many of which are already resistant to antibiotics, Martineau felt (and the FDA would eventually agree) that her experiment proved that eating Flavr Savr tomatoes would not increase antibiotic resistance among the microbes in our guts in a meaningful way.

After clearing the antibiotic resistance hurdle, Calgene turned to the Flavr Savr tomato itself. The tomato would have to be approved by both the FDA (to declare it safe to eat) and the USDA (to declare that it was not a plant pest). Calgene scientists began by describing all the ways in which the Flavr Savr tomato was intentionally different from other tomatoes. Flavr Savr tomatoes, for example, had an additional antisense copy of the PG gene and lower expression of PG compared to non-engineered tomatoes, both of which were straightforward to measure. Flavr Savr tomatoes also differed from other tomatoes in traits like improved fruit firmness, longer shelf life, and reduced susceptibility to post-harvest diseases, all consequences of the suppression of PG, and all easily measurable. More challenging, however, Calgene had to identify, measure, and report any unintended consequences of adding the antisense PG gene to the tomato genome. These might include reduced nutritional content or increased concentrations of glycoalkaloids (the toxic compound that accumulates in green potato skins). Unintended effects might manifest because of unanticipated interactions between the antisense PG gene and other genes, or if insertion of the Agrobacterium plasmid (or plasmids, as use of the gene gun meant that many copies may have gone into the same genome) disrupted another gene's function.

Because the chance of any unintended effect is smaller when fewer changes are made to the genome, Calgene wanted to find and focus on commercializing plants that only had one inserted copy of the antisense PG gene. Today's genome sequencing technologies would make this task simple, but these technologies were not available. Instead, Martineau developed a molecular assay that allowed her to count the number of inserts in each Flavr Savr tomato lineage. Her results revealed that of the original 960 plants into which the antisense PG gene had successfully integrated (fewer than 5 percent of seeds had any insertions at all), only eight had just one copy of the antisense PG gene. These eight became the tomato lineages on which Calgene staked its future.

Calgene grew tons (literally) of tomatoes from these eight lineages. Calgene scientists sent tomatoes to independent labs that measured their nutritional content and glycoalkaloid concentrations. One lab fed excessive amounts of pureed tomatoes to rats that were then harvested to look for signs of something going awry. Calgene held taste tests—although testers were not allowed to swallow the tomato or taste the locules (the cavities where the seeds, sugars, acids, and most of the flavors are) for fear that undigested seeds might make it into the environment. Calgene scientists performed firmness tests, dropping heavy weights on tomatoes and measuring how much the fruits compressed and poking tomatoes with sharp sticks to see how much pressure was required to break the skin. They measured the rate and timing of ripening on the vine and rates of post-harvest rot. The results were clear: Flavr Savr tomatoes were no different from traditionally bred tomatoes except that they rotted more slowly after harvest. Calgene submitted its petition to the FDA.

The FDA spent four years considering the safety of Flavr Savr tomatoes and antibiotic resistance genes. During this time Calgene complied with multiple requests from both the FDA and USDA for more information and additional experiments. Every day was, however, another day during which Calgene was spending but not making money, growing but not selling tomatoes.

One key question, however, was still unanswered: Would ripe Flavr Savr tomatoes remain sufficiently firm to stack and ship? Calgene's firmness tests were promising, but the shipping test would be the make-or-break moment. To perform this final and crucial test, Calgene scaled Flavr Savr tomato production in fields in Mexico. When the crop was ready, the Flavr Savr tomatoes were picked and piled into large bins. The bins were then packed into trucks and sent off on the 2,000-mile journey to Calgene's headquarters near Chicago, Illinois. Several days later, the trucks pulled into the Chicago parking lot with tomato sauce dripping out of their back gates. The results were in: ripe Flavr Savr tomatoes were the standard firmness of ripe tomatoes, which most of us know to pack carefully lest they become tomato mush. They were not the same firmness as green tomatoes.

Things never really improved for Calgene after the shipping fiasco. There was still value in tomatoes that were slow to rot, and Calgene continued to push for deregulation. There were good times, but mostly things just didn't go Calgene's way. Lacking a route to market, tomato growers withdrew from their contracts with Calgene to provide tomatoes for companies that could actually sell their fruits. Each time Calgene thought it was close to deregulation, the agencies asked for more data, which required more fruit, more experiments, and more time not selling tomatoes.

On May 18, 1994, the FDA released its official approval of Flavr Savr tomatoes and of antibiotic resistance genes, which the FDA decided to regulate as a food additive. The news was celebrated nationwide. The media (mostly) lauded the ripe-for-weeks tomato, and only one newspaper reported (incorrectly) that it contained fish DNA. Flavr Savr tomatoes were even reviewed positively by the Environmental Defense Fund, an organization that generally objects to biotechnology but felt that Calgene's voluntary subjection of its tomato to review provided confidence in the product's safety. Jeremy Rifkin's organization continued to object, mostly to the inclusion of the antibiotic resistance gene, but even his organized pickets, demonstrations, and tomato-squashing stunts could not crush the public's enthusiasm for the world's first genetically engineered whole food. In fact, the demand for Flavr Savr tomatoes outpaced supply to such an extent that grocers set daily limits on the number of tomatoes that each customer could buy.

The positive response to the Flavr Savr tomato was due to two decisions Calgene made early in the process. First, it volunteered the tomatoes for review by the FDA and the USDA despite not needing approval to go to market, knowing that regulatory review would cost the company time and money. This decision gave the public a chance to evaluate the risks associated with the engineered tomato. Second, Calgene did not try to hide the fact that the tomato was genetically engineered. Wherever Flavr Savr tomatoes were sold, the customer saw signs declaring proudly that this was a genetically engineered whole food. These signs included a brief description of the process and a toll-free number to call for more information. There was no opportunity for a customer to feel that they had been tricked into buying a product that they did not want to buy. The customer, and not just the farmer, had the power to make an informed choice about the Flavr Savr tomato.

Unfortunately, strong sales of approved-as-safe Flavr Savr tomatoes were not enough to save Calgene. As premium tomatoes, Flavr Savr tomatoes were the most expensive on the market, at around $2 per pound. Inefficiencies in growing and shipping meant, however, that Calgene was spending $10 per pound to get them on the grocers' shelves. Despite these challenges, the tomatoes were improving. New breeds with the Flavr Savr gene were both tastier and easier to ship. But a combination of bad growing seasons and mismanagement in the fields left Calgene struggling to get tomatoes to stores. In June 1995, just one year after they'd begun to sell Flavr Savr tomatoes, Calgene agreed to sell half of its company to Monsanto, which was interested not in the Flavr Savr tomato but in Calgene's plant genome engineering patents. Calgene tried to stay afloat with this investment from Monsanto, but it was too little and too late. In January 1997, Monsanto purchased the rest of Calgene's stock, and the company ceased to exist.

The Flavr Savr tomato did not fail because it was genetically engineered. It failed because of a series of poor business decisions, because Calgene knew too little about the fresh vegetable market, and because it was a small company (relative to the giants in the plant genetic engineering business) that elected to spend huge amounts of capital and time to pave the way for the future of biotech foods. Thanks in large part to Calgene's carefully designed, performed, and reported experiments, antisense gene technology and the antibiotic resistance gene that it used as a marker were ruled safe both in the United States and by the United Kingdom's Ministry of Agriculture, Fisheries, and Food. In laying the foundation for a new industry, Calgene was undeniably successful.

PRECISION GENE EDITING

As is clear from the Flavr Savr example, a key challenge of *Agrobacterium*-mediated genome engineering is that it is not possible using this approach to insert DNA at a predetermined place in the genome. Instead, the plasmid can end up anywhere: too far away from a gene with which it needs to interact to be effective, or inside a gene and disrupting something important. This problem has since been solved, thanks to newer technologies. In fact, three different technologies now allow precise, targeted genome editing. These technologies are known as programmable nucleases. Each can be synthesized in a lab to direct it to a designated place in the genome, where it will bind to and cut open the DNA strand.

Programmable nucleases work similarly to restriction enzymes. Restriction enzymes like *EcoRI* recognize a short sequence of DNA and then cut that DNA open. This makes it possible for scientists to splice together two organisms, stick new DNA into the genome, or make some other type of edit to the DNA. The problem with restriction enzymes is that the DNA sequence that they recognize is short, usually only a few letters long. A restriction enzyme delivered into a cell will chop the genome into many thousands of pieces, cutting everywhere that the few-letter DNA sequence that it recognizes is found. That's not good. The ideal DNA cutter for precision genome editing cuts the genome at only one specifiable place. Fortunately, this specificity increases along with the length of the sequence that the DNA cutter is programmed to recognize. If the recognition site is around 20 DNA letters or longer, this is often sufficient to match only one place in the genome.

The first DNA-cutting tool capable of achieving such sequence specificity was introduced in 1996. Zinc finger nucleases, or ZFNs, are made up of zinc finger proteins, which each recognize a sequence of three DNA letters, and a restriction enzyme (a DNA cutter) called Fok1 that has no sequence specificity (it'll cut anything). Zinc finger proteins were discovered in frogs but are found in most eukaryotic genomes, including our own, where their job is to bind to DNA in a way that changes the expression of a nearby gene. Once the recognition and binding mechanisms of zinc finger proteins were understood, scientists began to make new zinc fingers in the lab, tweaking them to bind to specific DNA triplets. Soon, scientists had developed a synthetic alphabet of zinc finger proteins that could be strung together to recognize long sequences of DNA. Paired with the restriction enzyme Fok1, ZFNs would find, bind, and cut DNA precisely as engineered.

ZFNs were the state-of-the-art, customizable, genome editing tool for nearly 15 years, but they are not perfect. They're expensive and cumbersome to design, requiring specialized equipment that most labs don't have. Their specificity is good but not great. Most ZFNs are programmed to recognize sequences of 18 DNA letters—nine letters on each side of where the cut is to be made. This is probably enough to match only one or a few places in the genome, but there is also a bit of wiggle room in the exact sequence that each zinc finger will recognize, which means it's hard to predict whether or how many secondary cuts will happen. Also, because not every DNA triplet has a zinc finger protein, some places in the genome are off-limits to this approach. Nonetheless, ZFNs were an enormous leap forward for genetic engineering and set the stage for the next generation of tools.

In 2010, transcription activator-like effector nucleases (TALENs) were added to the list of programmable DNA cutters. Like ZFNs, TALENs comprise arrays of molecules that recognize specific DNA letters and Fok1 to cut the DNA. Also like ZFNs, their role in the organisms in which they were discovered (*Xanthomonas* bacteria) is to bind to DNA and regulate the expression of nearby genes. Unlike ZFNs, however, the subcomponents of a TALEN recognize a single DNA letter rather than a triplet, making TALENS simpler to design. TALENs are, however, enormous molecules and difficult to deliver into the cell's nucleus. While scientists were working out that part, though, the third programmable nuclease entered the field and changed everything.

In 2012, teams led by Jennifer Doudna of the University of California, Berkeley, and Emmanuelle Charpentier, now director of a Max Planck Institute in Berlin, put the final pieces together on a gene editing system that some would say democratized genome engineering. The story began more than two decades earlier, when Yoshizumi Ishino and colleagues at Osaka University observed unusual arrays of repeated sequences in bacterial genomes. These repeated sequences, now known as clustered regularly interspaced short palindromic repeats, or CRISPR, are part of a system that evolved in bacteria to help them evade viral attack. Doudna and Charpentier discovered how to harness this system to edit genomes and, in 2020, won the Nobel Prize in chemistry for these discoveries.

In the CRISPR system, the palindromic repeats separate short fragments of DNA that match viruses that infected the bacteria in the past. Together with other molecules encoded by the bacterial genome (CRISPR-associated, or Cas, proteins), these short DNA fragments make up an adaptable bacterial immune system. Imagine an army of soldiers at the ready, each holding a flag that describes that soldier's target. The Cas proteins are the soldiers, and the flags are sequences between the repeats—the sequences that match infectious viruses. If a new virus with a sequence closely matching any flag invades the bacteria, the sequence (flag) will bind to that invading virus and the Cas protein (soldier) will chop it up, inactivating it.

Working with a Cas protein called Cas9 that evolved in *Streptococcus* bacteria, Doudna and Charpentier described how *any* sequence could be used as a flag to guide the Cas protein to a genomic target of interest. If, for example, I wanted to break the PG gene in the tomato genome, I could design a guide sequence (flag) based on what I know to be the sequence of the PG gene. After I deliver my guide sequence and Cas proteins into a cell, Cas9 would carry my guide sequence to the PG gene. The guide sequence would then bind to the matching PG gene sequence and Cas9 would cut the DNA, just as Fok1 cuts DNA in ZFNs and TALENs. Unlike ZFNs and TALENs, however, CRISPR guide sequences are cheap to make and easy to design since they're made of RNA letters rather than engineered protein complexes. They're also smaller and somewhat simpler to deliver into the cell. And they are more flexible. Since their original description, other Cas proteins have been discovered that do different things, such as cut only one strand of DNA or bind to the DNA and stay there without cutting, suppressing expression of the gene. With CRISPR-based gene editing tools, everyone can be a genome engineer.

Programmable DNA-cutting technologies changed the landscape of genetic engineering. Today, not only can DNA be inserted at a particular site in the genome, but single DNA letters can be swapped out and individual genes targeted, broken, and turned off. Precision gene editing has reduced the unintended effects of genetic engineering by minimizing the chance that cuts happen at more than one place in a genome. Remarkably, though, the only thing that these programmable nucleases do is find, and sometimes cut, the DNA. After the DNA is cut open, genetic engineers rely on different technologies to make sure that the edits they want to make actually happen. This does not always go exactly as planned.

ACCIDENTALLY TRANSGENIC

When the biotechnology firm Recombinetics set out to engineer a hornless Holstein, its intention was to make a single change in the genome—to replace one version of a gene with a different version. The intent was not to create something new. The polled allele, which causes the hornless phenotype, evolved in cattle centuries, if not millennia, ago. The firm knew that the polled allele could be bred into Holsteins using traditional approaches, but that this would lead to a deterioration in the quality of the Holstein breed as milk producers that would take generations to recover. Precision genetic engineering using programmable nuclease was the perfect solution. And, because the allele and its phenotype were well described and already part of the food chain, the FDA would almost certainly agree that the gene-edited Holsteins qualified for GRAS—generally regarded as safe—status.

This last point was a bit of a risk. When the experiment got underway in 2015, both the FDA and the USDA were still deciding what to do about this new category of engineered organisms: gene-edited organisms for which the changes to their genome could have occurred via traditional breeding. By definition, this category included only cisgenic organisms, since any movement of DNA between species (creating transgenic organisms) could not happen outside a laboratory setting. If Recombinetics' experiment went as intended, the hornless Holsteins would fall squarely into this category. Whether they would be acceptable to the agricultural industry would then depend on the agencies' decisions.

Recombinetics' scientists designed a TALEN that would bind to the region of chromosome 1 that contained the allele they wanted to replace. They delivered DNA that would code for the TALEN and the Angus version of the polled allele into cattle cell lines using a bacterial plasmid as a vector. What they hoped would happen next in each cell line is: (1) the cellular machinery would make a bunch of copies of the TALEN and the polled allele; (2) the TALENs would find both copies of chromosome 1 in the cell and bind to the matching sequences; and (3) the Fok1 part of the TALEN would cut the DNA on both copies of chromosome 1, creating a break in each. This damage to the DNA would then activate the cell's repair responses, of which there are two: nonhomologous end joining, in which the strands are simply stitched back together, often missing a base or so, and homologous recombination, in which the other copy of the chromosome is used as a template for repairing the DNA. The scientists needed the cell to choose homologous recombination. But instead of using the other copy of chromosome 1 (which they hoped was also broken by the TALEN) as a template to fix the damage, they wanted the cell to use the Angus version of the allele that they had delivered into the cell along with the TALEN. If all went well, both copies of chromosome 1 would be repaired in this precisely engineered way and nothing else would happen.

When the experiment was complete, the scientists screened 226 cell lines—the total number that they had attempted to modify—to check the results. Five had at least one polled allele, and three had the polled allele on both copies of chromosome 1. They then cloned the successfully edited cell lines, creating embryos that they hoped would develop into hornless bulls. Many months later, two healthy bull calves, Buri and Spotigy, were born.

Next, the scientists checked their work. Did any changes happen to the sequence of the polled allele during the experiment? Did TALENs bind to and insert the polled allele anywhere else in the genome other than the place on chromosome 1 where it was intended to go? The scientists sequenced the full genomes from the cell lines that had been used to create Buri and Spotigy and found that the polled allele had indeed replaced the non-polled allele on both chromosomes in both cell lines. They examined every site in the genome where the DNA sequence is close to that which the TALEN was programmed to find—61,751 sites in total—and found no evidence of additional insertions of the polled allele or parts of the polled allele. And neither Buri nor Spotigy grew horns.

Buri and Spotigy were eventually transferred from Recombinetics' farm in Minnesota to UC Davis, where Alison Van Eenennaam and her team continued to monitor their development. In 2016, Spotigy was sacrificed to analyze the quality of his meat, and Buri's semen was collected and cryopreserved. Alison used some of Buri's semen to artificially inseminate horned cows to test whether the polled phenotype would be passed to the next generation without unexpected effects. In January 2017, six pregnancies were confirmed. Then came the first major blow. The FDA released a decision that all gene-edited animals, regardless of whether their edits could have been achieved using traditional approaches, would be regarded as new animal drugs. With this decision, Buri and any of his descendants could not enter the food chain without a new animal drug approval—a lengthy and expensive process that neither Alison's team nor Recombinetics wanted to undertake. This was unexpected. Several months earlier the USDA decided to treat this category of editing as an accelerated form of normal breeding, and the FDA was expected to follow suit. Alison was disappointed but retained hope that data from what she anticipated would be perfectly healthy calves might encourage the FDA to reconsider.

In September, one female—Princess—and five male calves were born. All six were healthy, hornless, and, because of their classification by the FDA as drugs, destined to be incinerated. Alison's team performed physical exams, collected and analyzed blood, and sequenced each calf's DNA. When the calves were just over a year old, Alison sent these data to the FDA as part of an application to allow these animals to enter the food chain despite their classification as new animal drugs. Every indication was that the calves were entirely and unsurprisingly normal. Alison's team started to write up their results, highlighting the successes of the experiment that was nearing its end.

Then came the second blow. In March 2019, the FDA contacted Alison with bad news. Its scientists had gone back to Buri's genome sequencing data, which had been available to the public since 2016, and scanned it again. They looked for the polled allele (which was found right where it was supposed to be) and also for any DNA that matched the bacterial plasmid that Recombinetics used to deliver the gene editing tools into the cells. Unexpectedly, the bacterial plasmid was found right next to the polled allele. Presumably, the bacterial DNA had incorporated accidentally along with the polled allele during the gene editing process. Neither the team at Recombinetics nor Alison's team had seen this bit of bacterial DNA in Buri's genome. They hadn't looked for it. But the consequence was clear: Buri's genome contained both cattle DNA and bacterial DNA. He was, according to the rules, transgenic and would be regulated as such.

Now that Alison knew to look for the bacterial DNA, she re-screened the genomic data from her six calves to see if they, too, were transgenic animals. She found that four of the calves had the bacterial DNA next to their polled alleles. Two of the calves did not have bacterial DNA and so were not transgenic, although their father was. This meant that the mistake had happened on only one of Buri's chromosomes—the chromosome that these two cisgenic calves inherited. Princess, whom I would meet during my Halloween visit to Davis two and a half years later, had inherited the chromosome that contained the plasmid. Princess was accidentally transgenic.

As Alison revised her manuscript describing her calves, discussing their health and the successful passing down of the hornless phenotype without any untoward consequences to the animals, the FDA published its findings of bacterial DNA in Buri's genome on an online site that allows scientific papers to reach the public before they are reviewed. The press picked it up immediately. Sensational headlines disparaged Alison's and Recombinetics' work. "Gene-edited cattle have a major screwup in their DNA," wrote Antonio Regalado in the MIT *Technology Review*. Robby Berman pointed out "serious problems [in] gene-edited celebrity cows" in an article for the website *Big Think*. These articles pointed out another issue. The bacterial sequence in Buri's genome contained two antibiotic resistance marker genes. Neither could be expressed in cattle because the part of the bacterial genome that would make this possible was not included. That important detail was not, however, reported.

The media reports that followed the FDA publication were frightening, misleading, and damaging. In Brazil, regulators had been poised to collaborate with Recombinetics to create their own herd of polled dairy cattle, having decided, like the USDA, that this category of gene edits didn't require special oversight. Their plan was to evaluate Buri's offspring and, if all went as well as everyone assumed it would, create additional lines of polled dairy cows. They scrapped this plan when the news broke of bacterial DNA in Buri's genome. It didn't matter that Recombinetics had stopped using bacterial plasmids and adopted new approaches that deliver the editing machinery into cells using forms of DNA that cannot be accidentally incorporated into the host genome. It also didn't matter that Alison's results showed that the bacterial DNA did not affect the health of the animals, or that the bacterial DNA could be eliminated after one generation. The data didn't matter.

MORE DATA, BETTER EXPERIMENTS

When evaluating new technologies, it is important to separate facts from opinions. "GMOs are safe" is an opinion, just as "GMOs are unsafe" is an opinion. "Flavr Savr tomatoes had statistically no more or less vitamin C than the traditional variety to which they were compared" is a fact. Facts emerge from experiments designed to distinguish between competing hypotheses. If I feed laboratory rats a genetically engineered food, they will either get cancer more frequently than identical rats treated exactly as the others except that they are fed a non-GMO version of the food, or the two cohorts of rats will get cancer at equal rates. If the experiment is performed correctly, the result provides a new fact, which can be used to inform opinions.

Unfortunately, sometimes experiments are flawed. In 2012, Dr. Gilles Éric Séralini, a professor at the University of Caen Normandy in France, published the results of a rat study similar to the one I described above. He reported that rats fed herbicide-tolerant genetically engineered corn were more likely to get cancer than rats fed a diet of non–genetically engineered corn. Alarmed by these results, many in the media called for an immediate stoppage of production and sale of GM foods. This media narrative may, however, have been set in motion by Séralini. Normally, journalists confer with unaffiliated scientists before reporting about new results. In a break with these norms, Séralini gave journalists early access to his results only if they promised not to share them with other scientists until after his press conference. When the news broke, many in the scientific community cried foul, including scientists who supported labeling GM foods. They pointed out that Séralini used rats that nearly always got cancer over two years, which was the time frame of the study, and that he used too few of these rats for his results to be statistically meaningful. Séralini refused to release his data to scientists and governmental organizations that requested it, claiming that he was "being attacked in an extremely dishonest fashion by lobbies passing themselves off as the scientific community."*

Recently, a team of scientists funded by the European Union replicated Séralini's rat study. Lots of their cancer-prone rats got cancer, as in Séralini's study and as expected. But, having used five times more rats in each experimental group than Séralini used, the new study's statistical results were robust: Rats were equally likely to get cancer no matter what they ate. To date, Séralini's study stands out as the only one of many long-term and multigenerational studies to have found any difference between experimental and control groups of rats. Séralini's study was condemned by scientists and regulatory agencies around the world and eventually retracted. Yet it had the effect of establishing an anti-GMO scare story that, despite having been decisively falsified, persists today.

Sometimes experiments are not performed until it is too late. In 1999, more than a decade after the first field trials of Bt cotton, a team from Cornell University in New York reported in the journal *Nature* that windblown pollen from Bt corn threatened the survival of North America's beloved monarch butterflies. Bt evolved in bacteria to kill insects related to monarchs, so it should have been no surprise that their caterpillars fared poorly after eating leaves dusted with Bt pollen. Nonetheless, this finding was held up by anti-GMO activists as tangible proof of the environmental dangers of genetically engineered crops. Two years later, six larger field studies all concluded that Bt corn was in fact of little danger to monarch but-

---

* This quote, as well as other details of Séralini's now-retracted study, are from a report by Declan Butler, published on October 10, 2012, in *Nature*: www.nature.com/news/hyped-gm-maize-study-faces-growing-scrutiny-1.11566.

terflies, not because *Bt* isn't toxic to butterflies (it is) but because
the expression of *Bt* in the corn pollen is too low to be toxic and
because *Bt* corn is sparsely planted. These new studies did find that
one variety of *Bt* corn increased mortality of black swallowtails, another American butterfly. This variety has since been removed from
the market. If toxicity tests had been performed prior to approval
of these *Bt* plants, this mistake and the killing of black swallowtails
may have been avoided. Before condemning GM plants in favor of
traditional insecticides, however, remember that *Bt* sprays, which
are far more commonly used in agriculture than are genetically engineered *Bt* crops, are also toxic to monarchs and swallowtails.

Because experiments are sometimes poorly conceived or performed (or not performed at all), we don't always have all the facts
that we'd like to have to guide our decisions. Yet we still have to
make decisions and, in doing so, we accept explicitly some amount
of risk. How much risk we accept is both a personal and a situational
decision. My youngest child has been begging me for years to let him
play on a rope swing that hangs from a giant oak tree near our house.
To get on the swing he would have to climb a steep hill, clamber
onto the shoulder-height wooden plank, and then launch himself
over the path below. It's high and scary. I have no idea who made
the swing or how sturdy it is. My initial inclination was that it was
too risky. However, my son has grown into an athletic and relatively
risk-averse seven-year-old, and I've seen a few people much larger
and heavier than he on this swing, so I recently let him try it out.
While I don't regret my decision (much—we now have to make the
trip to the swing several times a week), I recognize that someone
else would have made a different decision based on the same facts.
The same is true when we make decisions about new technologies:
we vary in how much risk we are willing to accept, and we are capable
of changing our minds when new information becomes available.

The amount of risk that an individual is willing to take is also
situational. While a diabetic may have no qualms with GMO insulin

and a cancer patient might be willing to try an experimental GMO drug, these same people may not be willing to swap a traditional apple variety for one that is genetically engineered. The risk-to-reward ratio is different. For a sick person, the potential personal reward from a GMO drug may far outweigh the perceived risk. A healthy person choosing apples for a fruit salad may not value a good-looking salad sufficiently to outweigh whatever risks they associate with eating genetically engineered food. For GMO food products to be as widely accepted as GMO health products are, the biotech food industry must convince consumers both that its products are safe and that they provide genuine and novel value to the consumer.

What might this value look like? For some, knowing that genetically engineered foods can offset global food shortages, increase nutritional content, and allow farmers to use fewer chemical pesticides might be enough to outweigh whatever risk they associate with the foods. Others might need a more personal win. This might be that the product is less expensive thanks to reduced spoilage, increased yields, and lower costs to farmers. Improved aesthetic and flavor characteristics could also be important. I'd love to be able to buy tomatoes that stay firm for weeks and serve a fruit salad with apples that look and taste fresh even as next-day leftovers. I'd love these products, that is, as long as I know they're as safe to consume as their softening, browning counterparts. These risks can be evaluated and measured, just as Calgene measured the risks associated with eating antibiotic resistance genes. Of course, once those measurements are made, the consumer needs a way to find and digest that information. Today, this is difficult.

Golden Rice is probably the most famous example of a genetically modified food with a humanitarian motivation. Golden Rice contains two transgenes, one from daffodil and another from a soil bacterium, that make it express beta-carotene, the precursor to vitamin A, in the grains. If Golden Rice were to be widely adopted, it could help the more than 250,000 children who go blind every year because their diets lack sufficient vitamin A, half of whom die within a year of going blind. Despite Golden Rice's lifesaving potential, anti-GMO groups will say seemingly anything to dissuade people of its utility. Greenpeace claimed initially that Golden Rice didn't make enough beta-carotene to be useful. When the strains improved 20-fold, the Institute for Science and Society (another anti-GMO group) insisted that while this was still insufficient to solve the problem of malnutrition, it was also a toxicity risk. Other groups glommed on to the idea that beta-carotene was poisonous (it is not; our bodies excrete excess beta-carotene after making the vitamin A that we need) and used this as a call to reject Golden Rice while simultaneously insisting that people eat other plants rich in beta-carotene. Activists argued simultaneously that Golden Rice would fail because people wouldn't eat yellow rice (rice of all sorts of colors is eaten around the world), that Golden Rice should fail because it is a moneymaking scheme run by large Western businesses (the project leaders, who come from academic, governmental, and nonprofit organizations, intend to give away the seeds for free), and that Golden Rice would succeed to such an extent that it would be counterfeited, constituting a danger for people who thought they were getting real Golden Rice (and therefore vitamin A) but were not. The truth is that there is no argument too far-fetched or unrealistic, no conspiracy theory too absurd, and no experiment sufficiently convincing to change some minds, even when the intention is to save children's lives.

In 2008, scientists in China fed small servings of Golden Rice to 24 children to compare the uptake of vitamin A from Golden Rice to that from other beta-carotene-rich vegetables. They found that a single child's serving of Golden Rice provided 60 percent of a child's recommended daily amount of vitamin A—more than they received from a single serving of spinach. These experiments and data were reviewed by several independent committees, all of whom found the results reliable. In 2013 and after these results were made public, Greenpeace activists destroyed a field of Golden Rice in the Philippines, where 20 percent of children suffer from vitamin A deficiency.

There are small glimmers of hope for Golden Rice. In 2018, regulatory agencies in Canada, the United States, Australia, and New Zealand approved Golden Rice for cultivation and consumption by people and animals. In 2019, the Philippines approved Golden Rice for consumption.

While most organisms produced using the tools of synthetic biology are not engineered to solve humanitarian crises, many still have potential for considerable impact. In Uganda, a genetically modified banana that is both enriched in vitamin A and resistant to the bacterial wilt disease that has been devastating local crops was approved for cultivation at the end of 2017, promising a solution to the country's food crisis. In South Africa, scientists at the University of Cape Town are working to engineer the drought-induced dormancy trait that evolved in the resurrection plant, *Myroflammus flabellifolius*, into teff, a local indigenous grain, with the goal of making drought-tolerant teff. And in Hawai'i, the Rainbow papaya, a genetically modified papaya that expresses part of the coat protein of the ringspot virus, won regulatory approval, saved the local papaya industry, and perhaps converted a few GMO skeptics in the process.

Cisgenic GMOs, which are genetically engineered varieties that could have been engineered using traditional breeding, are becoming available in US markets. As of 2021, shoppers in the United States can buy non-browning button mushrooms and Arctic apples, both engineered to suppress the expression of the same browning gene and both deregulated by the USDA, which views them as no riskier to the environment than traditionally bred varieties of mushrooms or apples. While the non-browning trait may seem like a superficial change, nearly half the food grown in the United States is thrown away, and often for no reason other than that it looks unappetizing. Given that the United Nations projects that the world's farms will have to produce 70 percent more food by 2050 than they do today, a technology that stops us from wasting what we already have seems like a clear win.

While creating products that benefit consumers explicitly will no doubt improve the reputation of genetically engineered foods, consumers must also be able to evaluate risk. This means differentiating facts from the cacophony of lies and distorted half-truths about GMOs. There is no evidence that genetically engineered foods cause cancer, and no one has identified a mechanism by which this could happen. There are no genetically engineered vegetables on the market that contain genes from fish. And the bright green non-GMO label does not mean your food is free from new and uncharacterized mutations caused by the breeding process.

It is true that genetic engineering alters the genetic code—that is precisely the point of the technology. But it is wrong to lump together all genetically engineered products and all non-GMO products and claim that these two categories give consumers choice. Efforts to disambiguate categories of genetically modified products exist but are largely drowned out by frenetic anti-biotechnology propaganda.

There are signs that attitudes are changing. Anti-GM laws are weakening and sometimes being overturned in Africa and southern Asia. Clearer paths to regulatory approval are emerging in the United States. Some GM crops have won approval in the European Union. These changes hint that the present impasse may not last forever. A path to wider public acceptance of the tools of synthetic biology may also come from outside the agricultural realm, as similar technologies begin to restore ecosystem health and save species from extinction.

If and when that happens, it will have been too late for Princess and for her calf, who, as it turns out, did not inherit Buri's hornless allele. Princess gave birth on the last day of August 2020. Alison and Josie assessed her milk, looking for some effect of Princess's polled allele. As expected, they found nothing, as there's no reason that the polled allele would be expressed in milk. They added this new data point to their long list of things that are completely normal and expected from a Holstein, even an accidentally transgenic Holstein that never grew horns.

— CHAPTER 7 —

Intended Consequences

I ONCE ASKED AN AUDIENCE, "HOW MIGHT THE WORLD CHANGE if we brought mammoths back to life?"

Hands shot up.

"That would be the *sickest*!"

"We could see them and pet them and they would be all huge and hairy!"

"I could have one at my house, but it would probably be messy so maybe at the zoo is better. And we could even ride them!"

"They could live in Siberia and be friends with the reindeer, and the people who live there could eat them and make clothes from their fur!"

I should mention that the audience was my son's second grade class. Their teacher, Mrs. Jonas, had invited me to come talk about my career, and at this point was probably regretting it.

I'd been showing photos and videos from my mammoth fossil– hunting expeditions in Siberia, as an attempt to dissuade them that a scientist is always an old man wearing a white coat. I'd even brought a few chunks of mammoth tusk and bones to pass around the classroom.

As the mammoth bits moved from one table full of kids to the next, I felt the room start to turn. More hands came up, but hesitantly this time, and without any accompanying shouting.

"Maybe elephants would get mad because they would be jealous of all the attention mammoths got?"

Several heads nodded in agreement.

"Mammoths would probably want to live in the wild, but there's not enough people there to take care of them."

It was clear that their enthusiasm was beginning to wane.

"There would be hundreds and hundreds and they would knock down *all of the trees!*" worried a child dramatically, emphasizing the destructive power of hundreds of mammoths by ramming his clenched fist into the tower of glue sticks at the center of his table, toppling it.

The girl opposite him winced as a glue stick rolled toward her. She glared at him smugly and countered, "Not true because there *aren't* any trees in Siberia."

I was impressed. Not only had she refrained from throwing the glue stick at her exuberant classmate, but she'd also remembered a subtle detail of my presentation: the photos of our campsite on the treeless Arctic tundra.

"OK, but, would they have enough to eat then?" rebutted the glue stick toppler.

A student from the next table interrupted, trying to diffuse the situation. "Maybe they would get bored or starve and go extinct again."

More nodding heads.

After a moment of uncomfortable silence, another student spoke out, this time without raising her hand. "Maybe they can't have a nice life and we shouldn't make them come back."

Both I and the second graders fell quiet, contemplating the fate to which we had condemned these iconic beasts in only a few minutes of discussion: a second rise, this time human-made, and subsequent fall. A de-extinction, and a re-extinction. Because maybe there isn't a place for mammoths in today's world, as exciting as the idea of bringing them back might seem at first.

A bell rang, signaling time to go outside for recess. I stood by the door as they filed from the classroom into the sunshine, thanking them for listening and doing my best to collect the mammoth fossil bits that I'd passed around. They pushed past, some acknowledging me and others not, their attention now focused on more pressing tasks of being first to the monkey bars or securing the "king" position in foursquare.

As the last child left the classroom she paused and looked up at me. "If mammoths are gone because we killed them, do you think we have to give them another chance?"

I inhaled slowly, struggling for a thoughtful response. "I don't know," I said finally. "Do you think it would be fair to bring them back if the world that they lived in is gone?"

She shrugged noncommittally and looked at her toes. "It's just sad," she whispered at her feet as she headed out the door.

"Yes," I agreed, and then called after her as she ran toward the playground, "but we still have elephants!"

EXTINCTION IS FOREVER

Most of us working in the field of ancient DNA are accustomed to fielding questions about de-extinction—the term ascribed to the idea of using biotechnology to bring extinct species back to life. Have we done it yet? If not, how close are scientists to making it happen? Is it really possible to bring an extinct species back to life? How does de-extinction actually work? My answers are always the same. No, not yet, and probably not soon. It's not possible to bring back an identical replica of an extinct species, and it probably never will be. But there are technologies that may someday al-low us to resurrect components of extinct species—extinct traits. A scientist might, for example, modify an elephant by adding DNA that evolved in mammoths so that the elephant grows fur and thick layers of fat and can survive Arctic winters. Someone might modify a band-tailed pigeon so that its feather colors and tail shape make it look like a passenger pigeon. But will these modified elephants and band-tailed pigeons actually be mammoths and passenger pigeons? I don't believe so.

Why can't we bring back extinct species? There are a ton of reasons why long-extinct species are difficult to resurrect, ranging from technical hurdles that need to be overcome to ethical concerns about manipulating species to ecological challenges associated with releasing de-extinct species into a habitat from which they have been absent for perhaps tens of thousands of years. While some technical obstacles, like how to edit the germ line in birds or how to transfer a developing elephant embryo to a captive mom, might be overcome, others, like how to reinvent the gut microbiome of an extinct woolly rhinoceros or how to find a surrogate mom for a Steller's sea cow, are unlikely to be solved.

Let's consider mammoths. I know of three research groups currently working toward bringing mammoths back to life. Two of these, one led by Hwang Woo-suk from South Korea's Sooam Biotech Research Foundation and another led by Akira Iritani from Kindai University in Japan, want to resurrect mammoths by cloning—the process that most famously resulted in the birth of Dolly the sheep. Because cloning requires living cells, Hwang is hoping to find living mammoth cells preserved in the frozen mummies that are (thanks to global warming) melting from the Siberian permafrost. His institute is funding this work by charging people $100,000 to clone their pet dogs and using that income to pay the Russian mafia for access to newly discovered mammoths. The problem with this approach is that there are no living cells preserved in frozen mummies, as cellular decay begins immediately after death. Iritani's team acknowledges living mammoth cells are unlikely to be found and are instead turning to molecular biology to bring dead mammoth cells back to life, or at least close enough to life that they can be cloned. Iritani's plan is to coax proteins from mouse eggs that evolved to repair broken DNA in mouse cells to reconstitute the broken-up DNA in mammoth cells. In 2019, Iritani and his team published a scientific paper in which they attempted this with cells from a particularly well-preserved mammoth mummy called Yuka. Although the popular press heralded the paper as a sign of the inevitability of mammoth resurrection, the data seemed to indicate the opposite. Despite that Yuka's cells were in extremely good shape compared to cells from other mummified mammoths, the mouse proteins were unable to make much headway repairing the cell's DNA. Mammoths cannot be cloned because mammoth cells are all dead.

The third team hoping to resurrect mammoths, led by George Church from the Wyss Institute of Harvard University, accepts that living mammoth cells will not be found, given that the last mammoths died more than 3,000 years ago. But Church does not accept that this precludes reviving mammoths. He points to the ready existence of a limitless supply of living nearly mammoth cells—cells from Asian elephants—that can be grown in the lab and transformed from nearly mammoth to entirely mammoth using the tools of synthetic biology. To that end, Church initiated a program to use CRISPR to alter the DNA in Asian elephant cells a little bit at a time until the genome within the cell matches that of a mammoth.

Transforming an elephant genome into a mammoth genome is a daunting task. The lineages leading to Asian elephants and woolly mammoths diverged more than 5 million years ago. Because mammoth remains are well-preserved, scientists who work with ancient DNA have been able to reconstruct several complete mammoth genome sequences from well-preserved fossil remains. When these genomes were compared to the Asian elephant genome, the two genomes differed by around 1 million genetic changes. At present, it is impossible to make 1 million edits at once to a cell's DNA using any gene editing approach. Making lots of edits simultaneously means physically breaking the genome at lots of places simultaneously, which is a potentially disastrous situation from which the cell may not recover. Also, each edit or cluster of edits requires its own editing machinery, and delivering all of that into the cell at once is certain to go poorly. For now, Church's team is making one or a few edits at a time, checking that those edits are correct, and then using cells with the correct edits for the next round of edits. The last time I asked, he told me that his team had made around 50 edits to the elephant genome, swapping in the mammoth version of genes that research suggests may make a mammoth more mammoth-like than elephant-like. Today, Church's team has living cells that, if cloned, would contain the genetic instructions to resurrect some mammoth traits. Not mammoths, but something mammothish.

Could Church's mammothish cells be cloned? Cloning technologies, and in particular for domestic animals like sheep and cattle, have improved considerably since Dolly was cloned in 2003. It's taken longer, however, to figure out the necessary details—how and when to harvest eggs, how to create the ideal culture environment for early embryos, when to transfer embryos to a surrogate mom—for other species. The biggest barrier, however, is the reprogramming step, during which the somatic cell forgets how to be whatever type of cell it is and reverts to a type of cell that can become a whole animal. This step is so inefficient that cloning success rates rarely exceed 20 percent, even for those species that scientists regularly clone.

Elephants have never been cloned. One reason for this is that there is no market for cloned elephants. The market for clones of our domesticates is growing. The biotech firm Boyalife Genomics is building a cattle-cloning factory in Tianjin that it claims will eventually produce 1 million cloned Wagyu cattle every year to meet demand from the growing Chinese beef market. Hwang's Sooam

Biotech will clone your pet dog,* and ViaGen Pets, a firm based in Cedar Park, Texas, will clone your pet dog, your pet cat, and even your pet horse.† Few people, alternatively, are scrambling to clone their pet elephant.

It may not actually be possible to clone elephants. Elephants are large animals with equally large reproductive systems. This complicates key parts of the cloning process, like harvesting eggs for use in nuclear transfer and, because elephants regrow their hymen between pregnancies (the hymen has a tiny opening through which elephant sperm can pass but presents a significant and presumably crucial barrier to an elephant embryo), delivering a developing embryo into the uterus of the surrogate mom. Asian elephants are also endangered, which suggests that maybe the ideal use of this technology, should it become something that science can do, would be to make more elephants.

Even if elephant cloning does become technically (and ethically) feasible, it's not entirely clear that an elephant mom could carry a mammoth fetus to term. Five million years is a long evolutionary time, and 1 million changes is a lot of DNA differences. In fact, the evolutionary divergence between mammoths and Asian

---

* Hwang has also claimed to have cloned pigs, cows, and coyotes, and in 2004 published a paper in the journal *Science* in which he claimed to have cloned a human embryo. This latter claim has been proven false; the manuscript was retracted, and Hwang was indicted by the Seoul Central District Court for fabricating these data, embezzling research funds, and violating ethics laws by compelling junior members of his research team to donate their eggs for research.

† In fact, on August 6, 2020, scientists from ViaGen Equine announced the birth of a Przewalski's horse foal that had been cloned from 40-year-old frozen cells. The cells were part of the collection at the San Diego Frozen Zoo, and the project was a collaboration among ViaGen, the zoo, and Revive & Restore, a not-for-profit wildlife conservation organization that champions the use of biotechnology. Przewalski's horses are native to the Asian steppe and are severely endangered. Today, all surviving Przewalski's horses live in captivity and are descended from only 12 founders. The addition of this foal to the captive population will add welcome diversity to the founding gene pool.

elephants is about the same as that between humans and chimpanzees. It's hard to imagine a chimpanzee mother being able to carry a human baby to term or vice versa.

Some cross-species surrogacies have been successful, so evolutionary distance may not be a game-stopper. Domestic dogs have given birth to cloned wild gray wolf puppies, domestic cats to healthy cloned African wildcat kittens, and a domestic cow has given birth to a healthy cloned gaur calf. This research has revealed what everyone thought all along: the more distantly related the two species involved in cross-species cloning are, the lower the success rate will be at each stage of the cloning process. To date, the most distantly related species pair involved in a successful cross-species cloning experiment is dromedary and Bactrian camels, which diverged around 4 million years ago. Despite this long evolutionary time, a domestic dromedary camel gave birth to a cloned Bactrian camel calf in 2017. This is good news for Bactrian camels, which are one of the most endangered large mammal species alive today, and to conservation in general, as it highlights how advances in cloning technologies might expand the range of endangered species for which cloning is a viable tool in their recovery.

The only successfulish de-extinction to date also used cross-species cloning. In 2003 a Pyrenean ibex, also known as a bucardo, was born three years after her species went extinct. Four years earlier, a team led by Alberto Fernández-Arias, who was the head of the Hunting, Fishing, and Wetland Department of Aragon, Spain, harvested cells from Celia, the last living bucardo, and flash froze them so as not to damage their DNA. Fernández-Arias and his team worked over the next several years to develop a strategy to resurrect the bucardo. They tried to harvest eggs for cloning Celia's cells from another wild ibex, but the wild animals were unaccustomed to humans and great at escaping, and the experiment failed. Fortunately, they had better luck harvesting eggs from domestic goats. After replacing the domestic goat DNA in the eggs with Celia's DNA from her frozen somatic cells, they implanted 57 transformed eggs into surrogate does that were hybrids of domestic goats and Spanish ibex. Seven pregnancies established and one bucardo was born alive. Unfortunately, the cloned bucardo kid had a lung deformity, possibly caused by complexities of the cloning process, and survived for only a few minutes. Efforts to resurrect the bucardo using Celia's cells are on hold for now, but some of Celia's cells remain frozen.

While scientists may someday recode the elephant genome into a mammoth genome and clone that cell using an elephant mom, the developmental process itself could also be a technical barrier to resurrecting a mammoth. A cloned mammoth born to an elephant mom (or from an artificial womb, which is George Church's preferred solution to the problem of cloning elephants) would probably look like a mammoth. Everyone knows identical twins, and so we can all appreciate how influential DNA is in determining appearance. But our twin friends are not interchangeable. Their different experiences, stresses, diets, and environments shape two entirely different people. Would that baby mammoth, exposed to an elephant's prenatal developmental environment, raised by elephants, fed elephant foods, and hosting an elephant's gut microbiome, act like a mammoth, or would it act like an elephant?

Of course, none of this matters if the end goal is to create an elephant with a few mammoth-like traits, which perhaps is what we want to do. But if we're looking to create a mammoth, we need to re-create everything about the mammoth's environment from conception until death. That environment is, unfortunately, extinct.

SAME TECH, DIFFERENT GOAL

Why do we want mammoths back anyway? Some of us, like my son and his second grade classmates, just want to see them. Others want to study them, or to domesticate them, or to hunt them and eat them. Sergey Zimov, who directs northeast Siberia's Pleistocene

Park with his son Nikita, wants mammoths to transform the Siberian ecosystem and slow global warming.

Zimov's research has shown that when large mammals are present—Pleistocene Park is home to imported bison, muskoxen, horses, and reindeer—the grasslands that used to thrive in the region return and the permafrost warms less quickly. If that seems counterintuitive, consider what these animals do. They eat, wander around, and poop. This stirs up the soil, enriches it with nutrients from their manure, and spreads seeds. During the winter, the animals search for food, in some places clearing away the snow and exposing the bare ground to the freezing Arctic air, and in other places compacting the snow, creating a layer of dense ice on the soil surface. Sergey Zimov reports that winter soil temperatures are 15–20°C (27–36°F) colder in regions of the park with animals compared to regions where animals are absent. That means that animals reduce the rate of permafrost melting. Because more than 1,400 gigatons of carbon are currently trapped in the frozen Arctic soil (this is nearly twice the amount of carbon in Earth's atmosphere today), less melting permafrost means reduced release of greenhouse gases.

Elephants are the engineers of their ecosystems. They produce 100 kilometers (220 lb) of manure every day and transform the landscape by razing shrubs and (as the glue stick toppler suggested) knocking down trees as their herds pass through. Mammoths, the Zimovs anticipate, were similar ecosystem engineers. Because this role is currently unfilled in Pleistocene Park, the younger Zimov drives around in an old Soviet army tank, flattening the snow and toppling small trees. The Zimovs are hopeful that George Church's team will one day succeed, and that Pleistocene Park will be home to herds of wild mammoths doing their part to keep greenhouse gases trapped in the frozen soil.

While I can't confirm that mammoths would slow permafrost thaw, the Zimovs' rationale for mammoth de-extinction—to create a sustainable and diverse population of wild animals to fill a missing ecological role and thereby preserve an ecosystem—is to me the most compelling of reasons to bring back mammoths. But I wonder whether that requires actual mammoths, or if instead we might get the same benefit from Asian elephants engineered to thrive in Siberia. In other words, can we use de-extinction technology with the explicit goal to preserve habitats rather than to resurrect extinct species?

BIOTECHNOLOGY TO THE (GENETIC) RESCUE

Early in the morning of September 26, 1981, John and Lucille Hogg's dog, Shep, got into a fight. It was a brief fight, and Shep getting into fights was not unusual, so the Hoggs didn't bother to get up. When they did get up the next morning, John Hogg discovered the fight's outcome: a small animal was dead on his porch. He had never seen such an animal before. Its body was unusually long, with pale fur accented by four black feet, a black tail tip, and black tufts of fur in its large, pointed ears. It also had a small bright black nose and a stripe of black fur across its eyes that made it look as if it were wearing a mask. Hogg picked it up and studied it for a moment, and then chucked it into the bushes and headed back inside.

Later that morning, Hogg described the animal to his wife and children, its mask and black feet lodged in his memory. He thought it was probably some sort of mink, but maybe not. Lucille Hogg was intrigued, and they decided that, rather than throw the animal away, they should take it to a taxidermist in nearby Meeteetse, Wyoming, and have it mounted. The taxidermist would also probably be able to tell them what it was.

They were right. It took Larry LaFrenchie, the local taxidermist, just one glance at the creature's black feet to have a good guess at what it might be. With a concerned look on his face, LaFrenchie disappeared with Shep's trophy into the back room of his shop and made a phone call. Hogg remembers that when LaFrenchie emerged a few moments later he was visibly shaken and blurted out excitedly, "Oh my God, you've got yourself a black-footed ferret."* Hogg had never heard of a black-footed ferret, but he guessed from LaFrenchie's demeanor that he wasn't getting it back. He was right again. La-Frenchie had been told to confiscate the specimen and, as soon as he could, deliver it to officials at the US Fish and Wildlife Service.

Black-footed ferrets were part of the Class of 1967, the first list of species protected by the Endangered Species Act. Through the ESA, programs to save the black-footed ferret were initiated both in the wild and in captivity, but with little success. When Shep caught one, it had been more than seven years since a black-footed ferret was seen in the wild.

Black-footed ferrets are also known as prairie dog hunters, and hunting prairie dogs is what they do. Black-footed ferrets once ranged from Canada to Mexico, thriving anywhere that prairie dogs thrived. And as prairie dogs declined, so did black-footed ferrets.

Both of these declines are our fault. Prairie dogs live in prairie dog towns, which are clusters of deep burrows topped by mounds of loose sediment that can expand across more space than that occupied by many small towns. As European colonists and their descendants expanded westward during the early twentieth century, prairie dog towns disrupted farming and created obstacles to building and maintaining the roads and towns that supported the farms. There were also a lot of prairie dogs—perhaps billions at the turn of the twentieth century—which meant a lot of competition with cattle for food. To win that competition, settlers initiated widespread rodent-poisoning programs just as a new European disease—sylvatic plague—was mak-

---

* The Wyoming Game and Fish Department produced a series of free videos about the rediscovery and subsequent recovery of the black-footed ferret that includes interviews with John Hogg and the Hoggs' youngest daughter, Julie Sax, who was 10 years old at the time, and many of the conservation scientists who were involved with their project. These videos, along with additional information about black-footed ferret history and several black-footed ferret recovery programs, can be found at blackfootedferret.org.

ing its way through the prairie dog towns. Today, the rodent eradication programs have ended, but plague continues to infect and kill
prairie dogs. The population is now somewhere around 2 percent of
what it was at the turn of the twentieth century, although prairie dogs
remain widespread and genetically diverse. Some prairie dog populations have even begun to show signs of resistance to plague, which is
good news for their eventual recovery and long-term survival.

The outlook is less good for black-footed ferrets. Black-footed
ferrets fared poorly during prairie dog eradication, and they are also
susceptible to plague, which they catch by eating infected prairie dogs. By the end of the 1950s, biologists assumed that black-
footed ferrets were extinct. Then, in 1964, a small population of
black-footed ferrets was discovered in South Dakota near the small
town of White River. Biologists studied this population closely, hoping to learn what recovery approaches might work. As they followed
the ferrets, however, they noticed that many seemed ill. Believing
that they were running out of time, they captured nine ferrets and
brought them into captivity. Unfortunately, while several females
became impregnated and gave birth, no kits survived. The last wild
individual from the South Dakota population was seen in 1974, and
the last captive individual died in 1979.

Shep's discovery in 1981 of a black-footed ferret near Meeteetse,
Wyoming, provided an opportunity to learn what might have been
missing from the captive environment and again try to save them.
After months of tracking and trapping, biologists from the Wyoming Game and Fish Department and US Fish and Wildlife Service
estimated that the Meeteetse population had around 120 individuals, including many young ferrets, which they interpreted as a good
sign. Soon, though, things began to change. Rather than finding
healthy ferrets, the biologists were finding sick or dead ferrets. By
1985, it was clear that this population was on a path to extinction.
They made an emergency decision to capture as many ferrets as they
could and bring them into captivity.

Despite considerable effort, biologists captured a total of only 18 ferrets from the Wyoming population. Hoping to transform these 18 into a stable population, experts came together from the American Zoological Association, state and federal wildlife management groups, and the International Union for the Conservation of Nature's captive-breeding specialist group to map out a plan. Unfortunately, they still had little knowledge of what conditions helped black-footed ferrets breed in captivity. After months of effort and not much to show for it, biologists in the field captured an animal that they named Scarface. He would be the last black-footed ferret to join the captive colony. To their surprise and relief, Scarface bred with several females immediately, and his were the first surviving litters.

After Scarface, the future started to brighten for black-footed ferrets. Breeders discovered what worked and what didn't work, how to handle newborn kits, and how to prepare them for life after release. Today, black-footed ferret breeding programs are spread across the United States, with most of the world's breeding stock at five partner zoos. To minimize inbreeding, breeding decisions are made using pedigree information maintained in the black-footed ferret studbook. To make sure that as little as possible of the population's genetic diversity is lost, some females are inseminated using decades-old frozen sperm from founding members of the colony. Every year, around 200 black-footed ferrets are released as part of a collaboration of federal and state agencies, Native American and First Nations tribes, and private landowners. It is thought that the wild black-footed ferret population now numbers around 500 individuals, and that at least three of the 18 reintroduction sites are self-sustaining.

The black-footed ferret story is, for the most part, a story of successes. Their story highlights what can be won when experts come together to solve a hard problem. It shows that concerted efforts to find, study, and facilitate the recovery of an endangered species can bring that species back from the precipice of extinction.

It underscores the utility of conservation approaches like captive breeding, artificial insemination, and translocation. The problem is that despite these many successes, black-footed ferrets are still in trouble. While the breeding strategy has maintained most of the founding genetic diversity, all living black-footed ferrets trace their ancestry to seven individuals from a single population, so starting diversity was low. Once released, the isolated nature of their small populations leads to more inbreeding, further reducing genetic diversity. Black-footed ferrets in the wild also continue to be exposed to sylvatic plague and canine distemper. Although individuals can be vaccinated, a conservation program that requires capture and vaccination is not sustainable. Black-footed ferrets have been rescued from near-certain extinction, but we need new technologies if they are to be removed from the endangered species list.

Fortunately, these technologies may be in reach.

Drawing directly from the ideas on which de-extinction is based, it may be possible to inject new genetic diversity into the black-footed ferret population. As part of the San Diego Zoo's Institute for Conservation Research, the Frozen Zoo has been collecting and preserving tissues from threatened and endangered species since the 1970s. The Frozen Zoo collection, which is kept at a balmy 200°C below freezing, includes eggs, sperm, and embryos, as well as cultures of frozen cells that can be thawed, brought back to life, and regrown. These living cells preserve diversity from the past and are appropriate materials for cloning. The Frozen Zoo collection includes two black-footed ferrets: a male and a female that were collected and preserved by the Frozen Zoo's conservation biologists during the 1980s. Neither has any living descendants among today's black-footed ferrets.

Cloning new diversity into the population, however, may not be sufficient to save black-footed ferrets from extinction as long as sylvatic plague continues to infect them in the wild. Fortunately, there is another solution, and again an idea borrowed from de-extinction.

Just as scientists might add mammoth traits to Asian elephants to help them survive in cold places, plague resistance could be added to to black-footed ferrets to help them survive in places where plague is rampant. In this case, we don't need to look to extinct species for plague resistance genes. Instead, we can look to their evolutionary cousin, the domestic ferret, which is entirely resistant to plague. Or we can look to mice, a more distantly related species but one that also has evolved resistance to plague. Once we identify the genetic underpinnings of resistance, we can edit the black-footed ferret genome so that it, too, can be resistant to plague.

The technologies to edit the black-footed ferret genome exist. These are the same technologies that scientists have used to edit the genomes of crop plants and domestic animals to create disease resistance, herbicide tolerance, and hornlessness. The challenge is figuring out what part of the genome to edit. This won't be simple, but the path is understood. The genomes of domestic ferrets, mice, and black-footed ferrets will be sequenced and compared to identify where they differ. The genes involved with the immune system will be of particular interest, as these are likely to be involved in disease resistance. At the same time, scientists will study, using lab-based cell culture systems (rather than living animals), the process by which plague interacts with the cells in the immune system upon infection. The genetic analyses and lab-based studies will help narrow down the list of candidate genes—genes that may cause immunity to plague. Then scientists will walk through this list, using the tools of synthetic biology to alter the genome one candidate gene at a time, testing the efficacy of each edit until something works. Ultimately, this process will create a genetically engineered black-footed ferret that is identical in every way to today's captive-bred black-footed ferrets except that it will be immune to plague.

This work has already begun. In 2018, Revive & Restore, a non-profit whose goal is to support biotechnological solutions in conservation, received a permit from the US Fish and Wildlife Service to explore using synthetic biology to save black-footed ferrets from extinction. In collaboration with the San Diego Zoo, ViaGen, and several academic partners, Revive & Restore assessed whether the cell lines at the San Diego Zoo were viable for cloning, and began experiments to determine what genetic changes might provide heritable resistance to plague. In December 2020, a black-footed ferret kit was born—a clone of Willa, a female whose cells were preserved in 1983. Willa's clone, which they named Elizabeth Ann, increased the number of founders in the captive population to eight, and will provide a welcome and significant injection of genetic diversity. While this is only the first step in what will undoubtedly be a long process of scientific exploration, experimentation, and approval, it is a major win for black-footed ferrets and for genetic rescue. It is also the first step toward a sustainable population of disease-resistant black-footed ferrets prowling the American midcontinent for prairie dogs.

The intended consequences of genetically modifying black-footed ferrets are clear: to save them from becoming extinct by making them immune to the disease that is killing them. But what about the unintended consequences? What if, for example, the editing process or the edits themselves cause problems with gene expression or development? It is true that if errors occur during gene editing that break a gene or have some other bad effect on black-footed ferrets, these individuals will be weeded out of the population by natural selection. However, since plague is currently weeding black-footed ferrets out of existence, this would be no worse than what is happening now.

What if the modified genes escape into the environment? This question is important to consider in any project to release genetically engineered organisms into the wild, as the risk of genes escaping will depend on the organism's reproductive strategy. Some lineages coexist with closely related lineages with which they might interbreed, for example, which would provide a route for modified

DNA to enter a lineage other than that for which it was intended. The most closely related species to black-footed ferrets is the Siberian polecat. Siberian polecats can interbreed with black-footed ferrets but don't because their ranges are separated by the Bering Sea. Domestic ferrets are also closely related to black-footed ferrets and, while black-footed ferrets have not bred with domestic ferrets (at least as far as we know), this doesn't mean that they can't. If a domestic ferret escaped the comfort of its home and made it to the safety of a prairie dog town, it might have an opportunity to breed with a black-footed ferret. In the unlikely case that this happens, however, this is still not a likely avenue for genes to escape. Domestic ferrets probably have a suite of docility-related genes in their genomes that would make them or their hybrids fare poorly in the wild. Indeed, this is precisely why gene editing is preferable to normal breeding: gene editing transfers just resistance alleles, whereas breeding transfers everything, including an evolved fondness for belly scratches and head rubs.

What if solving one problem causes a different problem? For example, what if black-footed ferrets, suddenly released from their death sentence, become so numerous that they eat all the prairie dogs, upsetting the stability of their ecosystem? I doubt this is likely to be a problem, since black-footed ferrets already exist in this ecosystem and fill a remarkably strict ecological niche. Because black-footed ferrets prey almost exclusively on prairie dogs, the natural expansion of the black-footed ferret population will be limited by the size of the prairie dog population. And both populations will be kept in check by hawks, eagles, owls, badgers, coyotes, bobcats, and rattlesnakes, which will just as happily eat a black-footed ferret as they would a prairie dog.

To my mind, there seems to be little danger in releasing gene-edited, disease-resistant black-footed ferrets into their habitats, as soon as such an animal exists. So what is holding us back? A lack of basic genetic information, it turns out, is what is currently thwarting progress. Genomic resources are far less widely available for endangered species than they are for our domesticates. This includes reference genome sequences that can be used to detect inbreeding and maps that link particular genes or suites of genes to phenotypes. This makes it hard to identify putative resistance alleles, or indeed the underlying DNA that codes for any particular trait, in a threatened species.

I predict that, as genomic resources for endangered species become more widely available, synthetic biology will become increasingly important in conservation. We certainly have no shortage of problems to solve. Can synthetic biology transfer resistance to white-nose syndrome from European to American bats? Or help the world's coral reefs adapt to warming oceans? Or cure whatever it is that is killing Hawai'i's Ohi'a trees? Today, the answer is still no. But the technologies that will make these solutions possible exist, and molecular and conservation biologists, in collaboration with like-minded members of the public, are literally planting the seeds of the next, and to my mind extremely welcome, technological revolution in conservation.

THE TREE THAT REFUSED TO DIE

Despite that synthetic biology is being adopted more slowly in the conservation world than in the ag world, conservationists do get to claim one remarkable success story. And not only does this story include releasing a transgenic organism into parts of its native habitat, it's also a de-extinction story. Well, sort of, since the species was only functionally extinct, having hung on to life for nearly a century mostly in zombie form: belowground roots and rare, often small, ephemeral shoots. These living zombie trees, it turns out, are the perfect materials with which to begin a technological revolution.

At the turn of the twentieth century, the Appalachian forests of the eastern United States were dominated by the tall, broad, fast-growing, and prolific American chestnut tree, *Castanea dentata*. American chestnuts had been part of the landscape for hundreds of thousands of years, their massive bodies and seasonal seed crop providing room to some and board to others of the forest's squirrels, jays, wild turkeys, white-tailed deer, black bears, passenger pigeons, and dozens of other species, including people. And then, suddenly, the trees started to die. Small, burnt-orange spots appeared along the bark. These spots morphed into cankers, either swollen and cracked or sunken into pits as if the tree's flesh were rotting from within. The cankers grew, gripping the tree's girth like a tight belt and cutting off the flow of water and nutrients. The leaves above the canker withered and turned brown, and twigs and branches dried up and snapped off. Within a few months of the first spot appearing, the entire tree would be dead.

Soon after the trees began to die, William Murrill, curator of the New York Botanical Garden, identified the cause: a fungus known as *Cryphonectria parasitica*. The American chestnut trees in Murrill's gardens started showing signs of the disease around 1904, but scientists now believe the fungus entered the United States years earlier, hitching a ride in shipments of popular ornamental Japanese chestnuts that are resistant to the blight. Once the blight established in New York, it spread quickly. With each rain, tiny yellow tendrils emerged from infected trees and released millions of fungal spores capable of infecting neighboring trees. Within fifty years of the first tree death, all 4 billion American chestnut trees across their native range had succumbed to the fungus.

Although the grand American chestnut stands disappeared from eastern forests by the mid-twentieth century, 70 years later some trees are still not entirely dead. Belowground and out of reach of the fungus, root systems hang on, shooting up baby trees every so often that survive for a short time, rarely long enough to flower, before succumbing to the blight. Small stands of American chestnuts also survive in pockets of the American Midwest and Northwest, where they were planted by settlers in the nineteenth and early twentieth centuries. Unfortunately, even these isolated stands are now threatened. The largest surviving stand—a nearly 100-year-old cluster of American chestnut trees near West Salem, Wisconsin—began showing signs of fungal infection in 1987. The American chestnut is, in a filling-its-ecological-niche sense, extinct. But the zombie shoots and expat trees provide what every de-extinction scientist desperately wants: living cells.

Efforts to fill the vacant niche of the American chestnuts in eastern forests began in the 1920s. When imported fungus-resistant Chinese chestnuts failed to thrive in the American habitat, plant breeders tried to hybridize Chinese trees with surviving American trees—a standard approach in plant biotechnology. They hoped that breeding American and Chinese chestnuts would create offspring with both blight resistance and competitiveness in American forests. Unfortunately, while the resulting hybrids did inherit resistance to the fungal disease, they fared poorly, presumably because such a large portion of their genomes had evolved elsewhere.

In 1983, The American Chestnut Foundation was established, kicking off a decades-long quest to find a cure to chestnut blight. TACF was founded as a collaboration of conservation-minded laypeople and scientists, many from the State University of New York's College of Environmental Science and Forestry. Their first project was to improve hybrid trees by increasing the amount of American chestnut DNA in their genomes. TACF scientists embarked on a thirty-plus-year backcrossing program in which they bred resistant hybrids with pure American chestnuts. With each generation, the proportion of American chestnut DNA in each tree's genome increased. Today, after three generations, these trees are 85 percent American chestnut and 15 percent Chinese chestnut. However, they also have reduced resistance to the fungal blight compared to trees with more Chinese chestnut DNA, which suggests that resistance is achieved by a suite of genes working in concert rather than just a single gene. While imperfect, these trees are helping American chestnuts make a comeback. Today, TACF supports forty restoration stands of hybrid trees across the former native range of the American chestnut.

The American Chestnut Foundation's hybrid trees are good but not great replicas of purebred American chestnuts. The problem is that hybridization is imprecise and slow. It's difficult to control which parts of the genome are inherited, and the success of each generation can't be measured until the plant is sufficiently mature to show signs of fungal infection. Genomics can guide breeding selection, but this requires that breeders know precisely what part of the genome causes the resistance trait, which is not the case for American chestnuts. Of course, if the genes underlying resistance were known, one would not need to resort to the genomic messiness of hybridization and breeding selection. One could instead turn to genome engineering. Which is precisely what William Powell and Charles Maynard have been doing since 1990.

Together with TACF and an increasingly diverse team of specialists and laypeople, Bill Powell and Chuck Maynard, both professors at SUNY-ESF, developed the technological and biological know-how to create a transgenic American chestnut tree that is resistant to the fungal blight, and then they went ahead and did it. Their transgenic American chestnut tree is currently being evaluated by the US Environmental Protection Agency for release into the wild. If approved, it will be the first genetically engineered plant developed and used for the purposes of forest restoration.

To stop chestnut blight, Powell and Maynard needed to stop the fungus from spreading within the tree. As the cankers grow, feathered fungal fans spread into the plant tissue, producing a toxic compound called oxalic acid. The acid burns its way through plant cells to create holes into which the fungus moves. Powell and Maynard believed that if they could somehow neutralize the acid, then the fungus wouldn't spread and the tree wouldn't die.

Chestnut trees are not the only plant to have to deal with acid-producing fungi. Plants ranging from peanuts, bananas, and strawberries to mosses, grasses, and even other fungi all face similar threats, and many have evolved mechanisms to combat these threats. Powell and Maynard began by learning how these other plants dealt with acid-producing fungi. They transferred a few fungi-neutralizing genes into American chestnut genomes and saw that one gene—a wheat gene that produces an enzyme called oxalate oxidase—was particularly successful at conferring blight resistance on chestnut trees. Oxalate oxidase degrades oxalic acid and returns the pH of the plant tissue to nontoxic levels. Because this doesn't kill the fungus, it also reduces evolutionary pressure for the fungus to escape control. As a consequence, the fungus can coexist with American chestnuts, as it does with Chinese and Japanese chestnuts. There is other good news related to this particular transgene. Because it is a gene from a common food plant—wheat—and the enzyme that it produces—oxalate oxidase—is also produced by lots of agriculturally important plants, we eat a lot of it all the time. Should someone choose to enjoy a roasted transgenic American chestnut (which I intend to do as soon as possible), the proportion of oxalate oxidase in their diet would barely change.

The team's initial evaluations of transgenic American chestnut trees revealed more welcome news: trees that inherited only a single copy of the wheat oxalic acetate gene are resistant to blight. Because trees only need one copy of the transgene, transgenic trees can be outcrossed to wild-type trees, either purposefully in the lab or naturally in the wild, and a full half of their offspring will inherit resistance. All of the genetic diversity and local adaptations that survive in the zombie trees and expat populations can be bred easily into the transgenic population. The full diversity of the functionally extinct American chestnut population can be re-created without diluting their genomes—apart from the addition of one tiny gene that evolved in wheat.

The regulatory approval process for the blight-tolerant American chestnut tree has been complicated. As this is the first transgenic organism seeking approval for release into the wild rather than for use as a drug or on a farm, there is no regulatory precedent to follow. The trees are transgenic, which means the USDA's Animal and Plant Health Inspection Service has the option to regulate them as potential plant pests. In fact, all the work done so far, which includes experimental plantings in controlled environments (branches with fruits or flowers are wrapped in plastic bags to keep the transgenic cells in the confines of the experimental fields), has been done under permits granted by the USDA. The FDA could choose to regulate the trees as well, as the chestnuts will almost certainly be eaten by both people and domestic animals and the FDA's authority is to ensure the safety of our food. While the FDA has decided not to take the lead on regulatory approval of the blight-resistant chestnuts, it is reviewing the documents and experimental data submitted by the research team. The EPA also has regulatory authority over blight-tolerant American chestnuts through the Federal Insecticide, Fungicide, and Rodenticide Act because the added oxalic acetate qualifies as a "plant incorporated protectant," or PIP. And, because the native range of the American chestnut tree includes most of the eastern continent, the transgenic chestnut trees will also need to be approved by Canadian regulatory agencies, which Powell hopes will be expedited by his experience with the approval process in the United States.

In the meantime, the quest to restore the American chestnut tree continues. In 2019, transgenic, blight-tolerant American chestnut trees were planted at sites across several states, marking the start of a multistate, long-term ecological research project that will explore how eastern forests respond when American chestnut trees are restored. The TACF-and-SUNY team is filling nurseries with transgenic, blight-tolerant trees descended from the first outcrossing experiments in preparation for regulatory approval to distribute them to private properties, parks, arboretums, and botanical gardens. And outcrossing with wild-type trees continues to increase the genetic diversity of the resurrected population, which one day may be used to restore entire forest ecosystems.

Those who oppose the American chestnut restoration project worry about the unintended consequences of an experiment that might be hard if not impossible to undo. After all, American chestnuts have been functionally gone from these forests for nearly a century, and the forests may have adapted to their absence. If so, then returning American chestnuts may destabilize the current ecosystem in an unpredictable way. My own opinion is that the ecological risks in this specific case are few and far outweighed by intended rewards—a return to the ecosystem of the nutritional and structural benefits of a tree that never entirely disappeared.

The success of the American chestnut tree demonstrates the power of synthetic biology to help species adapt and survive, even if that means drawing evolutionary innovations from across the tree of life. But what if the conservation problem to solve is not that a species is threatened with extinction, but that its populations have expanded beyond our control? Could we use our new tools to engineer species in the opposite direction?

BLITZKRIEG REDUX

No one loves mosquitoes. They're small, they're quick, and they literally suck. My first exposure to Mosquitoes (with a capital "M") was my first field expedition to the Arctic. I was told the bugs would be bad, and I believed it. I grew up in the southeastern United States and have experienced evenings of "bad bugs." I brought bug spray. I wore long pants and a windbreaker. I felt prepared.

I was not prepared.

The first few days were fine. We were on the Ikpikpuk River in Alaska's North Slope. It was early July, which meant the raging waters that followed the spring melt had subsided into a gentle stream. We floated canoes lazily down the river, collecting the bones and teeth of ice age animals that, having been liberated from the frozen muck by the recently rushing water, littered the riverbanks and shallow sandbars. It was cool but not cold, the skies were clear, and the sun was out 24 hours a day. It was also pleasantly breezy.

Then the wind died down and the mosquitoes came out.

It was awful. They were little tiny things (the end-of-the-season dive-bombers, I was told; apparently the early season mosquitoes are larger and slower and easier to kill) and they seemed to start sucking my blood even before landing on my skin. No amount of poison repelled them. They appeared in clouds. I could squish sometimes dozens just by clapping my hands together in front of my face. If I hadn't been wearing a net over my head (a last-minute purchase from a store in Fairbanks), I'm certain that I would have been bitten up my nose and in my eyeballs.

We were not the only miserable animals on the river that day. With few warm-blooded creatures in the Arctic, every one of us who might improve a mosquito's brood size suffered. At one point, I saw a moose walking in the river ahead of us that was stopping every few minutes to submerge his entire body into the freezing water for a moment of relief. Even moose don't love mosquitoes.

My experience with mosquitoes in the Arctic was genuinely awful. But at least for now, Arctic mosquitoes have one thing going for them: They don't carry diseases. Their swarms made us miserable but did not put our lives at risk. This is not true of mosquito swarms in many other parts of the world, where a bite can be a death sentence.

Mosquito-borne diseases are one of the major threats to human and animal health worldwide. Mosquito bites transmit viruses, including dengue, West Nile, and yellow fever; parasites including those that cause malaria and lymphatic filariasis; and disease-causing bacteria. The World Health Organization estimates that every year 1 billion people are infected and 1 million people die from mosquito-

and other vector-borne diseases. Mosquito-borne diseases impact our food supply by infecting and killing livestock and are responsible for declines and even extinctions of species around the globe, in particular birds. We need to solve the problem of mosquito-borne diseases.

One option is to target the pathogens, but this has not been easy. Vaccines can prevent infection by some, but certainly not all, mosquito-borne diseases, and those vaccines that exist are often out of reach of the most affected communities. The modernization and globalization of our medical infrastructure have improved our ability to detect, track, and treat mosquito-borne pathogens, but newly emerging diseases like chikungunya and Zika virus challenge our capacity to keep up. What is needed is a way to target all the diseases at once. Instead of working to eliminate each disease separately, perhaps scientists should focus on discovering a way to eliminate the vectors—that is, to kill the mosquitoes.

Mosquito control is not a new idea. We've filled in swamps to remove mosquito breeding grounds, doused ourselves in toxic chemicals, and installed miniature electrocution devices (bug zappers) around our picnic tables. These measures work, but either inefficiently (zappers) or at a cost that we're unwilling to pay (remember when DDT was killing all the birds?).

Biology offers several environmentally friendlier options to control mosquito populations. Naturally occurring chemicals sourced from plants, for example, can kill mosquito eggs and larvae. Predators of mosquito larvae like fish, copepods, and frogs can be added to mosquito breeding grounds. Some countries have experimented with releasing bacteria and fungi that are pathogenic to mosquitoes. But these approaches also have downsides. Introduced species may outcompete native aquatic species, further disrupting the ecosystem. And because mosquito populations are so large (and natural selection therefore such a powerful evolutionary force), the chance that a new mutation will spread to make a population resistant to an introduced deterrent is high.

There is one biological approach to controlling mosquitoes, however, that is both ecologically friendly and resistance-proof. That approach is, counterintuitively, to release more mosquitoes. Not normal mosquitoes, though. Trojan horse mosquitoes armed with invisible superpowers that can destroy the mosquito population from within.

One trojan-horse-mosquito superpower is *Wolbachia*. *Wolbachia* are endosymbiotic bacteria—they live inside the cells of some insect species—and are passed from mother to offspring via infected eggs. Around 40 percent of insect species, including several mosquito species, are infected by *Wolbachia*. *Wolbachia* don't kill the insects that they infect but they do cause fertility problems. When an uninfected female mosquito breeds with an infected male, their offspring don't survive. In the late 1960s, scientists released large numbers of lab-reared *Wolbachia*-infected male mosquitoes into Myanmar (then Burma). These males bred with uninfected females, and no offspring were produced. The local mosquito population was eradicated, proving that this approach can work. *Wolbachia*-infected mosquitoes have since been introduced for biological control into mosquito populations in Australia, Vietnam, Indonesia, Brazil, and Colombia.

While promising, a few barriers stand in the way of widespread use of *Wolbachia* for mosquito control. First, it's difficult to produce only males in a laboratory environment. Because offspring of *Wolbachia*-infected females survive, the accidental release of *Wolbachia*-infected females along with males would allow *Wolbachia* to spread through the population, ruining its potential as a mosquito sterilizer. Second, any reduction of the mosquito population might not last very long if, for example, mosquitoes can easily recolonize from nearby. Finally, *Wolbachia* are already present in some of the most important disease-vector species, meaning that this approach simply won't work to control them.

The *Wolbachia* experiments grew from a theory that was initially developed in the 1930s. Raymond Bushland and Edward Knipling, both scientists at the USDA, had been tasked with finding a solution to the screwworm epidemic that was ravaging the country's cattle. They believed that they could get an upper hand on the disease by breaking the pest's reproductive cycle by overwhelming the population with infertile individuals. Borrowing from the same research that motivated mutation breeding in agriculture, they realized that zapping insects with X-rays would cause mutations to their DNA that would, with high doses of irradiation, make them sterile. After the end of World War II, their sterilize-then-release approach was tested in the field and was an enormous success. This approach has since been used to reduce diseases of cattle and other livestock, of agricultural crops, and of people. In 1992, Bushland and Knipling were awarded the World Food Prize for their role in developing the technique.

Like *Wolbachia*, sterilization by X-ray irradiation leaves no chemical trace in the environment and has limited direct effects on other species. Unlike *Wolbachia*, sterilization by X-ray irradiation could theoretically work in any species, and the only danger of releasing sterile females at the same time as sterile males is that money is wasted. The main drawback is the same as with mutation breeding. Irradiation induces random genetic changes across the genome, the consequences of which are unpredictable. High doses of X-rays may ensure infertility (and therefore ensure that other induced mutations don't spread) but may at the same time cause so many mutations that individuals are too weak or sick to reproduce. Alternatively, if the X-ray dose is too low, then irradiated individuals may remain sufficiently fertile to spread their X-ray-induced mutations into the pest population.

Fortunately, sterility can also be induced in insects via targeted gene editing. In 2013, Oxitec, a biotech company based in the United Kingdom, began releasing millions of genetically engineered, sterile, male *Aedes aegypti* mosquitoes into a suburb of Juazeiro in the Brazilian state of Bahia. To create sterile mosquitoes, Oxitec inserted a gene into the mosquito genome that causes the mosquito's cellular machinery to go into a runaway resource-draining death spiral. The gene makes a protein called a tetracycline repressible transactivator protein, or tTAV. During embryonic development, tTAV binds to the genomic element responsible for making itself, which causes more tTAV to be made. This tTAV then binds to that same element and makes more tTAV, and on and on and on, using up resources that otherwise would make proteins necessary for normal development. Edited embryos in mosquito eggs fail to develop into biting adults because of this broken cellular machinery.

But, the savvy reader might be thinking, if mosquito larvae die during development, how can adults capable of breeding and passing on their larvae-killing DNA be produced? Here, Oxitec scientists took advantage of a trick to the tTAV system: The expression of tTAV is suppressed in the presence of tetracycline, a common antibiotic. To produce adults, Oxitec rears its gene-edited mosquitoes in the lab in the presence of tetracycline. Then, the scientists sort the much smaller (and nonbiting) males from the larger females and release the males into the environment. When the released mosquitoes breed with wild females, their eggs contain embryos that inherit tTAV from their dad. Without access to tetracycline, these embryos die during development.

The 2013 release in Juazeiro was the second field release of Oxitec's engineered mosquitoes, which it named OX153A. The first trial had taken place a few years earlier in the community of West Bay, Grand Cayman Island. After only four months of releasing OX153A males in West Bay, the local *Aedes aegypti* population was reduced to just 20 percent of what it had been before the trial. In expanding to Brazil, Oxitec hoped to learn whether OX153A could be similarly successful in a more connected continental habitat. The scientists also wanted to establish local collaborations that would be crucial to optimizing the program for Brazilian habitats and, importantly, winning support of local communities and (eventually) regulators. The Juazeiro experiment was a success. After one year, the *Aedes aegypti* population in Juazeiro was reduced by 95 percent with no residual chemicals or toxins in the environment, and no direct effect on any species other than the invasive and deadly mosquito.

After two years of releases in Brazil, concerns were raised when a population-wide genome sequencing effort found evidence of small amounts of OX153A-like DNA in the genomes of Juazeiro's *Aedes aegypti*. This meant that some fertile OX153A mosquitoes were sneaking through Oxitec's sorting process and mixing their DNA into the wild population. This was expected. Sorting males from females by size is inexact, and Oxitec acknowledged prior to release that a few biting females were likely to be included in each release. Importantly, no transgenes (the tTAV gene or associated components) were found in the wild population, which indicated that the experiment was working correctly: OX153A mosquitoes that contained the transgene were not reproducing, and released mosquitoes that did manage to reproduce did not contain the transgene.

A second concern was that the population suppression won by releasing OX153A mosquitoes didn't last. Like other sterile insect approaches, successful long-term population suppression requires continued release. In Grand Cayman, the agreement between the local mosquito control board and Oxitec ended in December 2018. Within a few months, people living where the field trial had been taking place reported that the mosquito population was already significantly larger than during previous years. This, too, was as expected. The genetically engineered, sterile males don't survive for more than a few days, and they have to mate during that time in order to have any effect on the wild population. Once they die, there is no longer any possible impact from their release, and any gains in population reduction can be lost as mosquitoes colonize from nearby.

Oxitec already has a new lineage of transgenic mosquitoes that address these concerns. Its second-generation mosquito, OX5034, was announced in 2018 when the company withdrew a petition to the EPA to release OX153A in the Florida Keys, citing the newer and better mosquito in the pipeline. Oxitec's second-generation mosquito is similar to OX153A in that it also has tTAV, the deadly resource-consuming gene. However, in OX5034, the DNA construct with tTAV is inserted at a site in the genome where proteins are encoded differently depending on whether the mosquito develops as male or female. In females, tTAV is transcribed and the mosquito dies during development unless fed tetracycline. In males, tTAV is part of the genome but is not transcribed, and the mosquito develops normally. This new system—gene-edited females die but gene-edited males carry on as normal—has several benefits. First, males don't have to be separated from females in the laboratory prior to release. Instead, eggs can be placed directly into the wild to hatch and female eggs will simply not develop. Second, the sterility gene is heritable. Because males with the transgene survive, the transgene will continue to be passed down the male line. However, the sterility gene is self-limiting and will eventually disappear from the population.

The self-limiting aspect of the sterility gene works like this: Males that develop from OX5034 eggs have a copy of tTAV on both of their chromosomes. When they mate with wild females, all their offspring inherit one chromosome with tTAV. The female offspring will express tTAV and die, and the males will develop normally. When these males, which have one normal chromosome and one with tTAV, breed with wild females, half their offspring inherit tTAV. Of this half, the females die and the males develop normally. After ten or so generations during each of which the proportion of males in the population with tTAV is reduced by half, tTAV will disappear. Because the number of individuals carrying tTAV reduces in every generation, the population-reducing effect of self-limiting sterility declines over time. This strategy nevertheless has a much longer-term impact than one that requires repeated releases of sterile males.

In May 2019, Oxitec announced that its first yearlong trial release of OX5034 mosquitoes reduced *Aedes aegypti* populations by 96 percent in several densely populated neighborhoods of Indaiatuba, Brazil. On the heels of that success, Oxitec began working with other communities in Brazil to develop plans for additional releases of OX5034 mosquitoes. Oxitec is also developing plans to test the mosquito in locations outside Brazil. In September 2019, Oxitec petitioned the US EPA to release OX5034 mosquitoes at sites in Florida and Texas where local communities are showing increased support for a biotechnological solution to the problem of mosquito-borne diseases. In August 2020, in the wake of rising incidence of dengue virus infections in southern Florida, officials in the Florida Keys voted to allow the first release in the United States of OX5034 mosquitoes.

Oxitec's scientists are not the only group of synthetic biologists aiming to cure the world of mosquito-borne diseases. Target Malaria is a nonprofit collaboration among university scientists, stakeholders, bioethicists, and regulatory agencies whose goal is to eliminate malaria entirely. Target Malaria is focusing on strategies to combat *Anopheles gambiae* mosquitoes. Target Malaria is focusing on strategies to combat *Anopheles gambiae* mosquitoes, which are the main vector of malaria in sub-Saharan Africa. Target Malaria has teams in several countries where *Anopheles* mosquitoes are common and, in July 2019, completed the first-ever release of a genetically modified mosquito in Africa, in the village of Bana in Burkina Faso. The release was of sterile males, and a symbolic baby step. It demonstrated, however, both an excitement among local stakeholders to be involved in the science and the potential of the approach to reduce the local mosquito population.

Gene-edited insects can also reduce the impact of agricultural pests. In 2019, in collaboration with scientists from Cornell University, Oxitec released a self-limiting diamondback moth into an experimental field in upstate New York. Diamondback moths are major agricultural pests of brassica crops like cabbages, broccoli, and canola, and are known to rapidly evolve resistance to whatever pesticides are used to deter them. The gene-edited moth competed successfully with wild-type diamondback moths, and many fewer caterpillars were produced compared to control fields. Oxitec has also developed self-limiting strains of the fall armyworm, the soybean looper moth, and several other agricultural pests. The self-limiting sterility approach to reducing populations of crop pests could save farmers billions of dollars of losses globally every year while also reducing reliance on chemical pesticides. Intriguingly, it may also help shift the conversation around genetically engineered food, since genetically engineered crop pests (which people don't eat) could be used in place of genetically engineered, insect-resistant crops with similar gains in crop yield.

For the purposes of pest control, heritable sterility is an improvement over pest control strategies that rely on the continuous release of sterile males. And, because the sterility-inducing mutations disappear from the population over time, the natural ecological system remains intact. We meddle, but over the long term the impacts of our meddling are small.

For now.

A BETTER-THAN-RANDOM CHANCE

Our genomes contain nefarious elements: elements that subvert Mendel's Law of Segregation—the rule that every allele has an equal chance of making it into the next generation; elements that, because of their subversive power, sneak into the next generation more often than they should. Some of these elements warp the way chromosomes separate during meiosis (when cells divide to make sperm or eggs) so as to improve their chance of being in the cell that becomes the egg. Others, once part of a sperm or egg genome, destroy other sperm or egg cells. These elements are evolutionary villains, and they have names like "distorters" or "killers" or "ultra-selfish." They are also known as gene drives.

The elements that I described above are naturally occurring gene drives. But gene drives can also be created synthetically using the tools of genome engineering. To date, no synthetic gene drive has been released into the wild, but biotech companies, government agencies, and conservation biologists are all developing them. Gene drives are being contemplated for pest control, for invasive species management, and even to help species adapt to changing habitats. The appeal of synthetic gene drives is that they can spread a trait throughout an entire population more quickly than that same trait can spread by natural selection. This is also a reason to worry.

Target Malaria's plan to end malaria in sub-Saharan Africa includes three phases. Phase one includes the release of genetically engineered sterile males that began in Burkina Faso in 2019. Phase two is releasing a self-limiting version of *Anopheles gambiae* that, like Oxitec's self-limiting *Aedes*, can persist in the population for several generations before disappearing. In phase three Target Malaria plans to release a mosquito that will reduce the number of females all the way to zero. For this, the team will need to engineer a gene drive.

The gene drivers behind Target Malaria's audacious phase three plan are Austin Burt and Andrea Crisanti from Imperial College London. Since 2003, Burt and Crisanti have been working out how to construct a gene drive that can eradicate an entire mosquito population. With the invention of CRISPR-based gene editing in 2012, a solution came into focus. If the CRISPR components—the molecular machinery that finds and cuts the DNA in preparation for gene editing—could be included as part of the DNA edited into the genome, then the genome could essentially edit itself. The edit would be self-propagating.

For comparison, let's go back through what normally happens to edited DNA. In a normal genetic engineering scenario, the genome of an individual, let's say a male, is modified so that both of his chromosomes contain an edit. When he breeds with a wild-type female, their offspring will be heterozygous, meaning they will inherit one edited allele from their father and one wild-type allele from their mother. When these heterozygous individuals breed with wild-type individuals, half their offspring will inherit the edited allele and the other half will inherit the wild-type allele. This heritability pattern follows Mendel's Law of Segregation.

In a gene drive scenario, everyone inherits the edited allele. When an edited male breeds with a wild-type female, their offspring will initially be heterozygous; they will inherit the edited allele from their dad and the wild-type allele from their mom. But during the early stages of development, the CRISPR components in the edited allele will be transcribed—made by the cell—along with all the other proteins that the cell needs to function. Those CRISPR components will then find, cut, and edit the wild-type allele inherited from mom, transforming it into the edited allele. All offspring become homozygous for the edited allele. And since both chromosomes now have the edited allele (and CRISPR), the same thing will happen when those individuals mate with wild-type individuals. And the same thing will happen in the next generation. And the next. And so on. Eventually, every individual in the population will have two copies of the edited allele.

Given the above scenario, it's easy to imagine how a self-propagating gene drive can spread quickly. Maintaining that drive, however, requires that the drive allele remain intact. Any mutations that alter either the CRISPR components or the sequence of DNA that CRISPR is designed to recognize will break the gene drive. And if the trait being driven through the population reduces an individual's fitness, for example by making it sterile, there will be strong evolutionary pressure to break the drive. After all, sterility is, to put it mildly, evolutionarily disadvantageous.

To kill all the mosquitoes, Burt and Crisanti need an unbreakable drive.

To create an unbreakable drive, Burt and Crisanti chose to edit a gene called doublesex. In *Anopheles*, doublesex proteins are pieced together differently depending on whether a mosquito is male or female. Doublesex is also strongly evolutionarily conserved—any changes that occur to the sequence are likely to kill the developing mosquito. Because of this evolutionary conservation, doublesex is essentially unbreakable.

Seeing doublesex as the perfect target for an unbreakable gene drive, Crisanti's team decided to break the gene in a very careful way. It used CRISPR to make an edit that disrupts production of the female version of the doublesex protein. The edit has no impact on males, which develop normally. But females that inherit two copies of the edited allele are sterile. If introduced into a mosquito population as part of a gene drive, the doublesex mutation should eventually reduce the number of non-sterile females to zero.

In 2018, Crisanti's team released combinations of edited and unedited mosquitoes in small cages, and waited to find out whether their gene drive would work. All the caged populations went extinct within eleven generations.

Crisanti's team and Target Malaria still have lots of work to do before the phase three mosquitoes are ready for release into the wild. Experiments in larger field cages (and in still-closed but more natural habitats) will reveal how competition, predation, and other environmental factors affect the gene drive's success. More important, perhaps, will be the development of ethical and legal frameworks around the release of a potentially species-killing (or at least massively reducing) gene drive. Target Malaria is sharing their new technologies openly and including affected communities and other stakeholders when designing the plans for the next stages of their work. This approach allows them to build both the community infrastructure and trust of biotechnology in general (and gene drives in particular) that may someday allow the project to succeed.

BUT IT COULD ALL GO HORRIBLY WRONG

Kevin Esvelt is a professor at the Massachusetts Institute of Technology and will be the first person to tell you that gene drives are dangerous. Esvelt was the first person to figure out how CRISPR-based gene drives might work to spread a trait through a population, and he is certain that gene drives are the only near-term way to eradicate diseases like malaria. He has also been a leading voice calling for regulation of gene drive technology while simultaneously developing new gene drives in his own lab, although always engineered to fade away over time. Essentially, Esvelt is both excited and nervous. He also, as far as I can tell, never stops working.

Esvelt's research focuses on gene editing small mammals. He's developing, in consultation with local communities, systems to spread resistance to Lyme disease through white-footed mouse populations in Aotearoa/New England and to suppress invasive rodent populations in New Zealand. He is passionate about gene drive technology but aware of its risks. He is also adamant that gene drives should only ever be used if two things happen. First, the community in which the gene drive is to be used must understand and embrace the technology and the potential consequences of its use. Second, there has to be a way to turn the drive off. Oh, there is also a third thing. Esvelt insists that any scientist developing a gene drive should tell the world about their plans before they get started. Taking even the first step in developing a gene drive, he argues, is risking the future of this family of technologies and, more crucially, public trust in science. While Esvelt is confident that any gene drive can be stopped using molecular tools, he recognizes that damage caused by excluding affected people from the decision-making process may not be undoable.

Esvelt is serious about these criteria. He insists that not allowing people a voice about what happens in their environment is both foolish in a practical sense, since people often know a lot about their local environments, and also immoral, as what happens in their environments should be decided by them. His Lyme remediation project, Mice Against Ticks, is billed as a "community-guided effort to prevent tick-borne diseases by altering the shared environment." The Mice Against Ticks team is working with steering committees comprising community members from Martha's Vineyard and Nantucket Islands to decide what exactly is meant by "altering the shared environment." The problem to be solved is clear. Lyme disease rates are high on the islands—nearly 40 percent of people on Nantucket have had Lyme. The vector for Lyme disease is deer ticks, and humans get Lyme after being bitten by an infected tick. White-footed mice also get Lyme from infected tick bites and are the main sources of tick reinfection. Mice Against Ticks is looking to develop a strategy to, as the name implies, use the mice to target the ticks. The community remains divided, however, about whether that strategy should include releasing genetically engineered mice.

Mice Against Ticks has several potential solutions at its disposal, and it is up to the communities to decide how to proceed. Scientists could, for example, insert anti-Lyme antibodies into the white-footed mouse genome to make the white-footed mice immune to Lyme. They could release these engineered mice into the habitat and hope that they outcompete mice infected with Lyme. They could also engineer a gene drive to spread the anti-Lyme antibodies throughout the island mouse populations more quickly. For now, residents are more supportive of cisgenic than of transgenic solutions. This means they'd prefer that any DNA inserted into the white-footed mouse genome be sourced from white-footed mice, and preferably local white-footed mice.

The white-footed mouse populations on Martha's Vineyard and Nantucket Islands are relatively isolated from other mice. Given this isolation, a cisgenic approach might be sufficient to spread Lyme immunity throughout the populations, particularly if gene-edited mice were more fit than non-edited mice. Spreading immunity to mouse populations on the mainland, however, will require either more mice, more frequent introductions, or both, or perhaps the additional boost of a (necessarily transgenic, since it includes CRISPR) gene drive.

This raises a problem. What if a community on the mainland decides that they want to release transgenic white-footed mice with drives engineered into their genomes, but the communities on the islands do not? It might seem possible to stop gene-edited mice from hitching a ride on a ferry, but humans have a thousands-of-years-long history of inadvertently transporting rodents around the world. Indeed, this is precisely what caused many of today's invasive rodent problems that gene drives could be used to solve. If a mouse with a gene drive escaped from the mainland and ended up on Martha's Vineyard, there would be no barriers to it breeding with the local mice, and it's reasonable to expect that the immunity gene and associated CRISPR transgene would spread throughout the Martha's Vineyard population.

While the scenario above is bad because it overrides community choice, some conservationists worry that gene drives could have a different disastrous consequence. During the early phases of choosing how to rid Aotearoa/New Zealand of predators by 2050, an idea was floated to use a gene drive to extirpate invasive species by making them unable to reproduce. Ultimately, however, the community decided that it didn't want to use gene drives. One issue raised was potential escape. For example, the Australian-native common brushtail possum has become a major agricultural pest and conservation threat in Aotearoa and is on the list of invasive species slated for removal. If scientists were to engineer a common brushtail possum with a sterility gene drive, they could succeed in wiping them out across their range, with enormous benefits to Aotearoan native fauna. If just one escaped to Australia, however, the sterility drive could go into effect there as well, driving a native species toward extinction.

While I understand these concerns, I think the problem of escape could be countered with effective monitoring. Should a monitoring program discover that a sterility drive had invaded Australia, brushtail possums engineered to have resistance alleles to the gene drive could be introduced. As these would be more fit (able to reproduce), they would counter the escaped gene drive quickly and effectively. To ensure this safety check, however, Australian scientists would need to prepare by developing monitoring strategies and designing an effective counter-drive just in case. While this is entirely doable, the example highlights the need for teams considering gene drive systems to discuss their ideas early and openly with all potentially affected communities.

As we move forward as a society with gene drive technologies, we should assume that neighboring communities or governments will often make different choices about the release of genetically engineered organisms. We should assume that species will be difficult to contain within borders, in particular when borders lack physical barriers. We should assume that we will make mistakes, fail to anticipate an ecological impact, or change our minds. We should also acknowledge that there may be some situations in which invasive populations are sufficiently isolated that a rapidly acting gene drive is the most efficient and cost-effective approach to eliminating that population. Isolated islands with robust monitoring programs may meet these criteria, for example. But there will be other situations in which the possibility of connectivity cannot be excluded. In these situations, we should design our gene drives in such a way that we can counter them.

What might a gene drive kill switch look like? This is, as of today, an open question, but one for which many scientists are looking for answers. In 2017, the United States Defense Advanced Research Projects Agency (DARPA) announced that it would invest

$56 million in a Safe Genes program, with the goal to develop strategies to detect, reverse, or control gene drives. Kevin Esvelt, who was among the first to receive funding through this initiative, has a few ideas. One is to split the gene drive system into several drives that are scattered across the genome. These drives would function as a type of daisy chain, where the element at the base of the chain contains the instructions to drive the next element, which contains the instructions to drive the next element, and so on. While only the last element in the chain is necessary to express the genetically engineered trait, the other elements are what drives that trait to frequencies that are higher than by chance. Crucially, the element at the base of the chain is not driven, which means that, like tTAV in Oxitec's second-generation mosquitoes, it is self-limiting. Because only half the offspring born will have this basal, self-limiting element, it will gradually be lost from the population, ending the drive for the next element. Eventually, all the driving elements will be lost, and the population will return to as it was before the drive was introduced. Scientists can control how long an engineered trait persists in the population by adding links to the daisy chain or changing the number of edited individuals included in the release.

Daisy chains are just one of several ideas proposed to limit gene drives either spatially or temporally. New ideas will continue to emerge as the field expands, in particular as scientists, regulators, and stakeholders emphasize the need for caution.

While gene drives have been developed so far for only a handful of species, this is a fast-growing field. In 2018, Kim Cooper, a developmental biologist from the University of California, San Diego, provided the first evidence that a gene drive could work in a mammal. She inserted an extra gene into the mouse genome that activated CRISPR (which she also engineered into the mouse genome) at just the right time during germ cell development to make the engineered mouse genome edit itself—creating a gene drive. In this case, Cooper's intended change was to cause the mice to have entirely white fur. The results were not perfect. The drive worked only in females and 73 percent rather than 100 percent of offspring were white. But her team's success points to a future with a much wider range of organisms to shape into tools for biodiversity conservation.

Gene drives could also be developed for plants. These drives could spread genes that eliminate evolved resistance to herbicides. They could make a population resilient to an invasive pest or make a crop lineage better able to survive in an altered climate. As Kevin Esvelt would argue, these drives should all include either built-in mechanisms to limit their life span or established and tested approaches to counter the drive should it escape or have some impact other than what was anticipated. Such controls make gene drives local rather than global solutions for agriculture and conservation. They allow us to focus on achieving their intended consequences.

SOMETHING NEWISH

When we edit a species' genome, we are preempting evolution. We create an organism that never before existed. This also happens when organisms breed and their genomes combine: Their child is an organism that never before existed. When evolution creates something new, though, there is an element of chance in how those genomes will recombine. When we create something new, we overrule chance. We slightly tweak something that already was and we do so in a defined and specific way. We make something new—or perhaps newish—that is new in a way that we control entirely.

Our tinkering also has purpose. When we use a gene drive to propagate a trait through a population, we are using the organism that we created—the organism that never before existed—as a tool. That tool can reduce our disease burden, or clean up an ecosystem, or save a species from becoming extinct. Is it bad for us to manipulate species, to engineer them into something newish, for our own benefit? Is it justified to manipulate species if our intent is to benefit that species or to benefit the environment? Is it ethical for us to move genes between species as we shape them into new tools? Is it moral to look for solutions for our problems across the tree of life, even drawing from genomes of species that are no longer alive?

These are difficult questions to answer, but maybe an answer isn't necessary. We've been messing with the species around us for tens of thousands of years. For most of that time our tools have been dull, yet we've used them to create a world that is filled with beauty and opportunities and danger. Today, we are in a climate crisis, an extinction crisis, a hunger crisis, a trust crisis. We need all the tools at our disposal if we're going to survive these crises, and we need to learn how to talk openly and honestly about these tools. In fact, the tools that we have now are probably not enough to save us or other species or our habitats. For this, we need tools unlike any that exist today, tools that we have yet to imagine, because the road ahead is going to be bumpy.

But, for now, we still have elephants.

— CHAPTER 8 —

Turkish Delight

W E BEGAN THIS JOURNEY SOME 40 MILLION YEARS AGO,
when our earliest monkey-like ancestors colonized Africa
from Asia. A lot has changed since then. Continents moved, shifting the ocean currents around them. Global temperatures swung
from hot to freezing and back again. Habitats got wetter and drier
and transformed into something else entirely, creating opportunities
for plants, animals, fungi, and other microbes to evolve and diversify. And then, during the final 1 percent of the last 40 million years,
our own species appeared. Our ancestors spread across the planet
and became part—a key part—of the landscape within which species fought for space and resources. Some plants and animals and
microbes were good at living in a world that included people, and
they thrived. Others, lots of others, went extinct. Our species took
over the planet's habitats and changed them to suit our needs. Then,
during the most recent approximately 0.0005 percent of the last
40 million years, human society industrialized and reshaped Earth
nearly as much as the Chicxulub asteroid did when it slammed into
our planet roughly 66 million years earlier and ended the reign of
the dinosaurs.

That was the last 40 million years. If we now jump forward 40 million years, what will our Earth look like? Different than today, I expect, and not only because of us. No matter what we do, the continents will continue to move, and volcanoes will continue to erupt. In the next 40 million years, Africa will slam into Europe, Australia will merge with Southeast Asia, and California will slip up the coast of western North America into Alaska. The climate will be warmer 40 million years from today, and again not only because of us. Our sun, which is already a middle-aged star, is getting brighter. In around a billion years, our sun will be so bright and hot that our oceans will boil away. In the much more proximate future, 40 million years from now, our sun will be hotter than it is now, but the planet should still be habitable. But by whom? By us or lineages that descend from us? Or are we the last of our lineage, destined to fade away like the dinosaurs, making way for the Next Big Thing?

If we assume that our species will have a life span similar to that of other mammals, then we, too, are about halfway through our allotted time, with about half a million years still to go. We are, however, unlike other species. Other species become extinct because they're outcompeted, because they fail to adapt to a changing climate, or because they succumb to catastrophe. We outcompete all other species by killing them or taming them. We adapt to changing climates by engineering solutions outside of biology and, now, by engineering biology. We are susceptible to catastrophe but also wily, and perhaps we can engineer our way out of our own extinction as well.

As our powers to override evolution grow, so do our worries about abusing these powers. Where should we draw the line? Is editing plants OK, but editing animals off-limits? Is gene editing acceptable if it improves animal welfare or decreases pollution, but morally corrupt if the goal is aesthetic? What about modifying human genomes? If we do decide to edit ourselves, should we stick to making changes that affect only one person, or allow changes that pass to the next generation and alter permanently the trajectory of our evolution? What if we could cure a genetic disease, or protect our children during a pandemic? As we fret, habitats continue to deteriorate, species continue to disappear, new diseases continue to emerge, and people continue to starve and suffer.

When C. S. Lewis's Edmund Pevensie bit into the White Witch's Turkish delight, Pevensie did not suspect that the candy was enchanted. He knew only that it was more delicious than any candy he had ever tasted, and that he would do anything, even sell out his siblings, to have another bite. Will the ability to edit our own genomes be our Turkish delight? And if so, what will be the trigger that makes us take our first bite?

IMPOSSIBILITIES

I met Pat Brown for the first time when we were both guests of Google at its annual science-based "unconference" called Sci Foo. I spotted him in the kitchen holding court, sitting atop a tall counter stocked with plant-based snacks and surrounded by an audience of unconference attendees. He was wearing a white T-shirt featuring a cow's face stamped with a thick red prohibition symbol, gesticulating wildly as he mused, I imagine, about a future without cattle. I knew of Pat Brown both by reputation and because our lab had taken in a refugee from his lab when he decided to quit his Stanford job and start what would eventually become Impossible Foods. Unable to stop myself, I wandered up to the growing crowd to argue with him about cows.

Don't misunderstand. I love the Impossible Burger, which was the first product that Impossible Foods released. The Impossible Burger is genuinely tasty and, in my opinion, much more similar to beef burgers in both texture and flavor than other currently available plant-based options. This is of course what Pat Brown intends: to make a meat substitute for people who eat meat. The Impossible Burger does not trouble me. What troubles me is Brown's wish to get rid of all cattle, which I don't see as necessary. To my mind, even those who abhor the beef and dairy industry should see a place for cattle in our world. They are, after all, descendants of aurochs. If nothing else, cattle could at the least live where aurochs once did and fill the role of a large grazing herbivore in a rewilded natural ecosystem.

Pat Brown spent 25 years as a professor of biochemistry at Stanford University before he decided to switch careers. In 2009, he took an 18-month sabbatical to figure out what to do next. He'd already invented the DNA microarray, which changed the way scientists catalogue differences between DNA sequences and in the quantities of proteins that genes make. He'd also cofounded the Public Library of Science, which changed the way scientists publish their research by making their work free to access for anyone anywhere in the world. For his next project, he wanted to instigate change on a bigger scale. After some research and contemplation, he decided to change the way people eat by removing all animal products from the planet's food system.

Pat Brown may be a little bit crazy, but he's definitely a lot smart. He knew that asking people to stop eating meat would never work. Animal foods provide nutrition and satisfaction that plant-based foods are missing, and animal foods are tightly woven into many cultures. To remove animals from our diets entirely, Brown would have to invent a plant-based product that mimicked exactly the animal products that people want to eat. He turned to his biochemical and technology development savvy and founded two companies: Impossible Foods, which focuses on plant-based meat replacements, and Lyrical Foods, which develops plant-based dairy products like cream cheeses, yogurts, and filled pastas and sells them under the brand Kite Hill.

Impossible meat's success was impossible to fathom when the product was first introduced. In 2009, most people assumed that Brown would find some success in a niche market. In 2016, the Impossible

Burger debuted to unexpected fanfare at New York City's Momo-fuku, an award-winning, meat-centered bistro run by Chef David Chang. In 2019, Burger King introduced the Impossible Whopper, and in 2020, the Impossible Croissan'wich, made with Impossible pork. Today, national restaurant chains and local eateries in several countries are serving Impossible Bolognese, Impossible tacos, and pizzas topped with Impossible sausage. Impossible meat can be found on grocery store shelves and in the meat section at big-box stores. The secret to Impossible Foods' success was its key ingredient, which, in addition to giving Impossible meat its subtle "moreishness," could be grown in vats.

Brown's team discovered that what gives beef its flavor and mouthfeel and even color is a molecule called heme. Heme is a component of hemoglobin, the blood molecule that carries oxygen from our lungs to all our other cells. Heme binds iron, which is why our blood (and a rare steak) has a slight metallic flavor. All living things have heme, but some organisms have more than others, and the more heme that something has, the more complex its flavor profile is. Pat and his team discovered that the trick to making a delicious meaty(ish) burger is to pack it full of heme.

Plant heme, of course.

Plant heme is identical in molecular structure to animal heme but present in much smaller amounts per unit of plant mass. Among the most heme-rich of plant parts are the roots of legumes like beans, where heme (which is also red in plants) takes part in the nitrogen fixation process. Soybean roots are particularly heme-rich for a plant and a good option for a plant-based source of heme for Impossible meats. However, growing millions of acres of soybeans to harvest their blood (heme) is not the environmental win that Brown envisioned. Instead, he needed something scalable, preferably vertically. The solution, fortunately, had already been invented, the model in fact coming from the very first product of synthetic biology: human insulin.

Yeasts are tiny protein factories. Yeasts are fast-growing, single-celled organisms that are easy to keep alive and, unlike bacteria, can make proteins that are ready to use with little additional processing. Yeasts are also straightforward to engineer using recombinant DNA technologies. Since the 1980s, when yeast was first used to produce recombinant insulin, protein production using yeast has grown into a billion-dollar market. The process is simple. The scientist inserts the gene that codes for the desired protein into the yeast genome, and then places the genetically engineered yeast into a fermentation chamber (a giant vat similar to those used to make beer). The yeast eats water and sugar and multiplies, making more yeast and, at the same time, lots of the protein that the yeast genome has been engineered to express. In the final step, the scientist purifies the protein from the soupy yeast mixture and does whatever they want to do with the purified protein.

Impossible Foods realized that if it could engineer yeast to express the hemoglobin proteins that evolved in soybean, then it could produce soybean heme in massive quantities. Even better, it could do so in a vertically scalable way. And this is exactly how Impossible Foods manufactures its not-so-secret ingredient: plant blood, biosynthesized by genetically engineered yeast, in vats. The company combines the purified heme with other ingredients like soy, sunflower, and coconut oils, as well as seasonings and binders that together make Impossible meats look, cook, bleed, feel, and taste like minced meat. The products have about the same amount of protein as animal meat, with fewer calories, a lot more sodium, and slightly less fat.

The secret to the Impossible Burger's success is that, thanks to the heme, the patty is a pretty good—and improving—approximation of a beef burger. It's not trying to be a tasty veggie patty or a super-healthy meat substitute. It's trying to be beef, but without the cattle.

Meat and dairy are not the only animal products that biotech companies are using yeast to produce. Bolt Threads, for example, engineered yeast to produce spider silk proteins that can be spun into fibers and knitted into fabrics. Modern Meadow engineered yeast to produce collagen, the protein that makes skin stretchy but tough. The purified yeast-produced collagen is pressed into sheets that are then tanned, dyed, and sewn into whatever products traditional leather might be used to produce, like bags or briefcases or even furniture. The US Department of Energy's Joint BioEnergy Institute engineered microbes that produce indigoidine, a synthetic molecule used to dye denim. Lonza, a biotech company that creates products for the pharmaceutical industry, engineered yeast to produce a protein used to confirm that medicines are free of toxic microbes. This protein, recombinant factor C, functions identically to, and can therefore replace for this purpose, a protein that most of the industry sources from the blue blood of horseshoe crabs, killing thousands of crabs every year. And Ginkgo Bioworks is engineering yeast to produce flavor and fragrance compounds normally isolated from fields of herbs or flowers. These bioengineered compounds allow companies to avoid business pitfalls like bad weather or failed crops and free up agricultural land for other purposes.

Yeast is also not the only organism that can be put to work as a molecular factory. Plants, too, can be engineered to express genes and make proteins in ways that benefit us rather than them. For example, Surinder Singh from Australia's Commonwealth Scientific and Industrial Research Organisation (CSIRO) leads a team of scientists who use genetic engineering to alter the compositions and yields of oils extracted from plant seeds, stems, and leaves. One of Singh's goals is to engineer plants to produce oils that are sufficiently stable at high temperatures to replace petroleum-based oils as industrial lubricants. Another is to engineer canola, a common oilseed plant, to express a gene from algae that produces long-chain omega-3 fatty acids. Demand for these fatty acids for aquaculture and nutritional supplements has depleted the oceans of algae-eating fish like pilchards and anchovies from which omega-3 fatty acids are normally sourced, with ripple effects up the ocean food chain. Harvesting fatty acids from plants instead provides an environmentally sustainable source of this important healthy oil.

The ability to make proteins with only a DNA sequence and an engineerable biological factory also allows experimentation. A few years ago, my lab worked with Ginkgo Bioworks and artist and smell researcher Sissel Tolaas to create a fragrance from extinct flowers. My lab extracted and sequenced DNA from dried flowers of the Maui hau kuahiwi, which was indigenous to Maui, Hawai'i, and last seen in 1912; the Falls-of-the-Ohio scurfpea, which was last seen near Louisville, Kentucky, in the 1920s; and the Wynberg conebush, which was native to Cape Town, South Africa, and last seen in the 1800s. We then isolated fragrance-producing genes from our ancient DNA extracts and sent these sequences to Ginkgo, which engineered these sequences into yeast. After fermentation and purification of the fragrance compounds, Tolaas composed a smell that became part of an immersive traveling art installation into which visitors were invited to enter and breathe in an aroma that they would never otherwise experience: a genetically engineered fragrance made from three flowers that have been extinct for over a century.

What's most exciting to me about the extinct flower project is not that we resurrected extinct fragrances, which is pretty cool, but that our goal was something other than mimicry. We were not trying to re-create a copy of something that was as good as possible, but instead were using biology and evolution and engineering to make something new, something perhaps better than what nature designed. Yes, the compounds that Tolaas used to compose the new fragrance were engineered by nature and reengineered by us. But the final product was entirely human designed. And when people walked through the exhibit, they experienced both the past and the future simultaneously: a small taste, or in this case smell, of what our biotechnologies can do.

Could creating rather than replicating be the next phase of genetically engineered foods and other products? I imagine synthetic biologists of the future racing not to create the most beefy plant-burger or the most sausagey plant-patty or even a perfume that best approximates the smell of a flower that used to be alive, but to create something more delicious and more wonderful than anything we've yet imagined. In our experiment with Gingko, we borrowed the design of our fragrance compounds from nature, but we didn't have to. We could instead have created our own fragrance genes by stringing together amino acids that our research or instinct suggested would result in an interesting smell. We could have expressed our synthetic proteins in yeast and given the result a whiff to decide whether it was great or could still use some tweaking. Free from most evolutionary constraints, we could have mixed and matched modified, synthetic fragrance molecules to create a smell to satisfy any craving.

With synthetic biology, we no longer have to remain within the bounds of what we can imagine. And this makes it hard to predict what new tools or solutions or products we might create as the field advances.

RED FISH, BLUE FISH, GREEN FISH, NEW FISH

In 2002, the California Fish and Game Commission banned aquarium fish breeders from selling danios—the small, normally navy-and-white-striped tropical freshwater fish that pet shops label as "easy" for beginner fish keepers. Normally, such a ban would arise from concern that the animal in question had potential to become invasive. Not so in this case. Danios are tropical fish and ill-prepared to survive in California's frigid waterways, and, although danios have been kept in tanks for decades, no free-ranging population has ever been discovered in California. These particular danios were highly unlikely to establish local populations, as they essentially advertise their presence to predators. In fact, this was precisely the trait that most concerned the California commissioners: the danios glowed.

The pet pariahs in question were GloFish, a lineage of genetically engineered danio that, at the time, was newly for sale in the United States. GloFish had been developed a few years earlier in Zhiyuan Gong's research lab at the National University of Singapore as part of a project to engineer a fish that would be capable of alerting bystanders of contaminated water. Gong chose danios because they were relatively straightforward to engineer compared to other fish. Danio eggs have a clear outer membrane that makes developing embryos visible from the single-cell stage. If edits can be made to the genome at this earliest developmental stage, then not only will all the fish's tissues be engineered in an identical way, but the edits will also be passed to the next generation.

Gong's goal was to engineer a danio that could be a living sensor, sending out a visible warning signal if it swam into polluted water. To achieve this, he first needed to identify genes that are expressed only in the presence of pollutants. He found these genes by exposing danios to toxins like estrogen and heavy metals and measuring which genes were expressed. He also needed to engineer a visible cue that would be expressed at the same time as these pollutant-sensing genes, so as to alert bystanders that the pollutant was present. For this, he turned to a gene that evolved in jellyfish called green fluorescent protein, or GFP. When expressed, GFP makes a protein that absorbs UV light from the sun and emits it as lower-energy green light. Gong's plan was to insert GFP close to the pollutant-sensing genes in such a way that their expression became paired. When a genetically engineered danio swam into polluted water, it would express both genes and light up like a fluorescent green lightbulb.

To prove that they could engineer danios to glow green, Gong's team inserted GFP into the danio genome without linking it to any other gene's expression. The engineered danios glowed! And Gong was not the only one to notice. Within two years, Alan Blake and Richard Crockett, two businessmen from Austin, Texas, secured a license from the National University of Singapore to use Gong's technology for a different application entirely. They would make and sell glowing fish as pets.

Blake and Crockett formed a company called Yorktown Technologies, which took to market two glowing danios produced by Gong. One was an "electric green" danio with the color induced by GFP, and the other a "starfire red" danio engineered to express a red fluorescent protein gene that evolved in corals. Blake and Crockett also went to work developing fish in new and brighter colors and adapting the technology to other common aquarium fish species. Today, fish keepers who live where GloFish haven't been banned can stock their tanks with five different glowing fish species—danios, bettas, barbs, tetras, and rainbow sharks—that come in colors like "sunburst orange," "galactic purple," "cosmic blue," and "moonrise pink." Each fish's genetically encoded dazzle is derived from genes that evolved mostly in corals and sea anemones. In 2017, Yorktown Technologies sold the GloFish brand to Spectrum Brands Holdings for $50 million. At the time of sale, they estimated that GloFish accounted for around 15 percent of the United States market share of aquarium fish sales.

While glowing fishes are the only genetically engineered pet that is (relatively) widely available for purchase today, glowing cats, dogs, rabbits, birds, and piglets also exist, although these were all developed as scientific tools rather than as pets. Since its discovery, GFP has become a popular marker gene, replacing antibiotic resistance as a means to confirm experimental success—that the primary edit was successfully made to the genome. For example, scientists at The Roslin Institute in Scotland created transgenic chickens whose genomes contain both GFP and an edit that makes the chickens resistant to bird flu. Using GFP as a marker, they could track simultaneously which chickens had been edited successfully (they glowed under ultraviolet light) and the incidence and spread of bird flu (they were sick), allowing them to test whether their edit protected chickens from getting bird flu.

Glowing technology was also used to track success while engineering the first transgenic dog, Ruppy, short for Ruby Puppy. Ruppy was born in South Korea in 2009, one of a litter of four cloned beagles engineered by scientists at Seoul National University to express a red fluorescent protein gene. The experiment was a proof of concept; the team only intended to show that transgenic dogs could be cloned. Ruppy and her genetically identical littermates looked like perfectly normal beagles under natural light. But under ultraviolet light, they all glowed a charming, bright, ruby red. When Ruppy was mated to a non-transgenic dog, half her puppies inherited the red protein gene, indicating that the transgene had incorporated successfully into her germ line.

Although not yet available for adoption at your local animal rescue, genetically engineered pets like Ruppy are in our future. Some might glow under ultraviolet light, but the majority will be manipulated to express traits that improve our pets in some substantive way. As more of us move into urban settings, for example, we might want miniaturized versions of pets that are better adapted for apartment living. In fact, the Beijing Genomics Institute, or BGI, had exactly this in mind when it announced in 2014 that it would soon begin selling a genetically engineered micro-pig: a Bama pig genetically engineered to stop growing at an apartment-friendly size of around 14 kilogram (30 lb), about 25–35 percent of the size of an unedited Bama pig. BGI nixed its micro-pig mass production plan a few years later without explanation, but rumors spread of supposed micro-pigs that, while they grew slowly, nonetheless eventually reached a normal not-so-apartment-friendly size.

Future engineered pet breeds will probably be better versions of themselves. Synthetic biology will allow us to maximize the traits for which each breed was initially selected—hypoallergenicity, hunting prowess, exceptional olfaction—but without the sloppiness of breeding. Some of this work has already begun. Scientists from the Guangzhou Institutes of Biomedicine and Health announced in 2015 that they created a double-muscled beagle by knocking out the beagle's myostatin gene, which is the same gene responsible for the double-muscled phenotype in Belgian Blue cattle. While the Guangzhou team purports that the extra-beefy beagle will be particularly useful in police and military contexts, I'm not convinced. If they can make labs and spaniels better at sniffing out cancers, though, I'll be 100 percent on board.

The tools of synthetic biology will also be used to make our pets healthier and to improve our relationships with them. Once scientists have figured out which mutations cause Dalmatians' susceptibility to bladder stones and boxers' predisposition to heart disease, gene editing will remove these maladaptive variants from the breeds altogether. As understanding grows of which genes map to which traits, synthetic biologists can use these data to create, for example, cats that don't express allergens in their saliva and golden retrievers that don't shed.

With synthetic biology, however, we need not be limited to traits that already exist. What new pets might we create once we reach outside our traditional selective breeding toolbox? No longer limited by evolutionary constraints, traits could be combined across species barriers, and even across time. We might make dogs that chirp and birds that meow, saber-toothed house cats and woolly guinea pigs. We might make beagles with wings and terriers that lay eggs, and glowing fish that can climb out of the tank and walk across the room for a cuddle. Obviously, these are all fantastical creatures, some patently absurd, and none that science can even come close to making today. We know far too little about how genes interact and what combinations and timings of gene expressions are responsible for most of these traits, and there is no path to combining complex traits that evolved over long and separate evolutionary trajectories. Still, I caution against ruling out even the craziest of ideas. Imagine how our hunter-gatherer ancestors would have laughed at the suggestion that people would someday transform wolves into chihuahuas.

Or, for that matter, that we would transform the rotting remains of dead plants into a storage vessel that would never, ever, ever degrade.

THE GREAT GALACTIC GARBAGE PATCH

There is not actually a trash island twice the size of the state of Texas floating in the eastern Pacific Ocean. This often-quoted idea originated when Charles Moore, a racing boat captain and oceanographer, who while on a return trip to California after a sailing race between Los Angeles and Hawai'i, found his boat surrounded by a cloud of floating plastic debris. Confused and concerned, Moore began to transect the area to see if he could discover the source of the plastic. When he called in his report, he noted that the debris was strewn across an area forming an island twice the size of Texas. This simile hooked the media and has been used ever since to describe what soon would be known as the Great Pacific Garbage Patch.

The Great Pacific Garbage Patch is enormous. Some estimates suggest it spans more than 1.6 million square kilometers (more than 600,000 mi^2; about twice the size of Texas or three times the size of France), and others suggest it might be much larger. It is not, however, an island of garbage across which one might be able to walk. Some ocean trash explorers have found a few conglomerations of larger bits of plastic—baskets, buckets, fishing nets, sneakers, candy wrappers, toothbrushes, soap bottles—but the garbage patch comprises mostly tiny broken-down bits of microplastic that look like large grains of pepper floating in a giant bowl of ocean soup.

The Great Pacific Garbage Patch is also not the only ocean garbage patch, it's just the most famous of them. Five ocean garbage patches—two in the Pacific Ocean, one in the Indian Ocean, and two in the Atlantic Ocean—accumulate where rotating ocean currents called gyres suck in and aggregate trash as it floats across the ocean.

The ocean garbage patches are also not really patches at all. Rather than accumulate as floating rafts of trash at the ocean surface, the garbage patches transmogrify. Bits break off and float away with the current or sink through the water column to settle on the ocean floor. The patches themselves move with the winds and currents, invading marine communities as they near and then retreat from continental shores.

In addition to being gross, the ocean garbage patches are harmful. Marine mammals, large fish, and turtles get caught in the tangled mess of discarded fishing nets and other bits of garbage in a phenomenon that has been dubbed "ghost fishing." Birds mistake small balls of plastic and styrofoam as fish eggs and feed them to their young, which die of starvation or from internal injuries. Chemicals absorbed by the plastic—most other materials biodegrade so the garbage in the patches is mainly plastic—are consumed by ocean life and transmitted up the food chain all the way to us, with unknown effects on their and our health. Human industries ranging from tourism to fishing to shipping are impacted as marine ecosystem health declines, suffocating in discarded plastic.

Plastic pollution is not going away by itself. Unlike food or wood or cotton scraps, bacteria don't eat plastics and so plastics do not biodegrade. This is not particularly surprising, since plastics didn't exist until people made them and so there has been no long-term evolutionary pressure for bacteria to evolve to break plastics down. Instead, plastics photodegrade—UV light breaks the bonds that hold together the long chains of inorganic molecules. How long that takes depends on where the plastic ends up. In a landfill (or deep in the water column), plastic buried under meters of other trash (or water) will have little exposure to UV light and will degrade slowly, perhaps taking more than 400 years to break down. Although plastic floating at or near the ocean surface will break down more quickly, plastics disintegrate rather than dissolve or disappear, eventually becoming the microplastic soups that are our ocean's garbage patches. In the meantime, people continue to dump around 8 million tons of plastic into the world's oceans every year. And our dependence on plastics for storage, packaging, clothing, and other products continues to grow.

Recently, chemical engineers have begun to tweak synthetic polymers to create partially or even entirely biodegradable plastics. Adding starch during polymer synthesis, for example, makes a somewhat biodegradable plastic. While the starch may encourage microbial degradation, however, it also changes the properties of the final product and may speed up disintegration into microplastics. Plant-based plastics made from compounds like cornstarch are starting to be used in some contexts, for example as disposable forks, disposable cups (for holding cold beverages—they melt at relatively low temperatures), and disposable packing materials. Plant-based plastics are often said to be compostable, even edible. Their chemical composition is similar, however, to that of petroleum-based plastics, which makes this claim a stretch at best and sometimes an outright lie. Some plant-based plastics will eventually biodegrade in industrial composting facilities, but they will persist in backyard composting bins and in landfills just as petroleum-based plastics would. Worse, plant-based plastics cannot be recycled alongside petroleum-based plastics. If mixed accidentally with petroleum-based plastics bound for recycling, the entire batch becomes contaminated and unrecyclable.

A newly discovered class of plastic materials called polyhydroxy-alkanoates, or PHAs, are a promising alternative to both petroleum-based and plant-based plastics. PHAs are produced by some bacteria naturally as a strategy to store energy when resources are scarce. In a factory, these bacteria can be induced to produce large amounts of PHAs by restricting some nutrients and providing an overabundance of others. Unlike petroleum-based or plant-based plastics, PHAs biodegrade in backyard compost bins, in terrestrial landfills, and even in the ocean. PHAs also have wide potential utility; so far scientists have discovered more than 150 different PHAs that microbes make from staples like sugars, starches, and oils. These PHAs can be used in isolation or blended with other materials to produce biodegradable plastics with broad durability, malleability, and heat and water tolerance properties.

While global industrial production of PHAs is still small, biotechnology firms are developing PHA polymers that can replace many of the nonbiodegradable plastic products that inevitably end up in the environment. Existing PHA-based products include capsules for slow-release agricultural fertilizers, microbeads for facial scrubs, microplastics that improve the UV-blocking properties of sunscreens, grocery packaging for fruits and vegetables, and single-use cups and cutlery for fast food chains. The largest market for PHAs today is mulching film—the plastic barrier that farmers place over cultivatable land to prevent weeds from penetrating crops. Mulching film is often plowed back into the soil at the end of each growing season, where it disintegrates into microplastics that stick around for hundreds of years. Mulching film made of PHAs will instead biodegrade.

Microbial production of PHAs also presents new opportunities for synthetic biology. Today, bacteria produce PHAs on industrial scales by metabolizing sugars and vegetable oils. But as scientists learn more about what regulates and constrains these microbial pathways, they will be able to engineer microbes to make PHAs from different starting materials. They could, for example, engineer microbes to produce bioplastics from waste products like wort left over after beer production, coffee grounds, or garden scraps.

Or even clean up petrochemical oil spills or biodegrade petroleum-based plastics.

As we look to the future, I have little doubt that engineered microbes will someday exist that not only make the products that sustain our way of life but that also help clean up our planetary garbage. This solution may come sooner than later. Recently, scientists discovered microbes that are capable of breaking down some petroleum-based plastics. The natural rate at which these microbes consume plastics is too slow to make much of a dent on the present-day pollution problem. Genetic tinkering will, however, optimize these processes. Who knows, microbial engineers may even discover how to harness the energy released when these microbes break the bonds that hold synthetic polymers together. Synthetic biology could literally turn one era's trash into another era's treasure.

SAVE OUR SOILS

Today's pollution problem may have been an inevitable consequence of our species' evolutionary success. In the last 200 years, the number of humans living on this planet grew from 1 billion to nearly 8 billion. We all eat, need a place to sleep, and create waste, both organic and inorganic, that has to go somewhere. Plastic pollution is part of the problem but certainly not the only environmental challenge that synthetic biology could be harnessed to solve. The global industrialization of goods production and agriculture have polluted our planet's air and water and degraded arable land. The consequences are dire; the United Nations estimates that farming production will need to increase by 50 percent to feed the projected 9 billion people who will inhabit the planet in 2050, yet the University of Sheffield's Grantham Centre for Sustainable Futures estimates that today's global capacity to grow crops is a third lower than it was 50 years ago. The degradation of arable land comes in part from seasonal tilling to remove weeds, which releases the soil's carbon into the atmosphere, contributing to the accumulation of greenhouse gases. It also reduces soils to their mineral components, which makes them vulnerable to drying up and washing away, contributing to eutrophication in the rivers and oceans. As farmers work to produce crops from soils that are less naturally fertile, misuse or overuse of herbicides and pesticides exacerbates the problem by changing the mineral composition and pH of the soil and destabilizing the community of microorganisms that maintain healthy soil ecosystems.

Synthetic biology has already improved crop production and slowed the deterioration of land under cultivation. Genetically engineered herbicide-tolerant plants reduce the need for tilling by allowing farmers to control weeds using herbicides like glyphosate and glufosinate.* Genetically engineered, insect-resistant plants reduce the impact of chemical pesticides on biodiversity and soil quality. Genetic engineering has created varieties of crop plants that are resistant to disease (e.g., Rainbow papaya), more nutritious (e.g., Golden Rice), more attractive (e.g., Arctic apples), and better able to thrive in less than ideal growing conditions (e.g., flood-resistant rice). Tomatoes

---

* Glyphosate, which is the most popular herbicide used worldwide, has received considerable scrutiny since 2015, when the International Agency for Research on Cancer (IARC) released a statement that categorized glyphosate as a "probable carcinogen." This agency evaluates carcinogenicity differently than other agencies in that it assigns carcinogenicity based on whether a substance could possibly cause cancer under any conditions, regardless of whether those conditions are ever likely to occur in the real world. Other major health organizations consider carcinogenicity risk based on realistic exposure scenarios, including measuring exposure duration and concentration. In decades of trials both before and after the IARC report, no major health or regulatory organization, which includes the US EPA, the World Health Organization, the European Chemicals Agency, Health Canada, and many others, found glyphosate exposure to increase cancer risk to people. The IARC, however, considers glyphosate among group 2A "probable carcinogens," which also includes working night shifts, drinking hot beverages, wood smoke, working as a hairdresser, and getting malaria. This IARC classification is the study on which recent civil law cases have been based.

have also diversified and improved thanks to gene editing technologies. Zach Lippman, a geneticist from the Cold Spring Harbor Laboratory in New York, tweaked three tomato genes using CRISPR to create a cherry tomato variety in which the fruit grows in bunches like grapes on a vine and matures quickly. This more compact and productive cherry tomato is intended for small spaces like urban rooftop gardens or, perhaps, gardens grown in a human colony on Mars.

Our domesticated animals, too, will increasingly benefit from the tools of synthetic biology, and not only from improvements to the nutritional content of their feed. Genetic engineering will, I am hopeful, eventually be allowed to both improve animal welfare on the farm and boost production of food products. Signs of acceptance of animal genetic engineering are emerging. AquAdvantage salmon grow at twice the rate of regular salmon and thereby halve the time that it takes to get a fish to market. AquAdvantage salmon was approved for sale (but not to be grown) by the US Food and Drug Administration in 2015. In 2019, new leadership of the FDA removed the prohibition of importation of AquAdvantage salmon from Canada, where it had been approved for both production and sale, and green-lighted production of AquAdvantage salmon within the United States. In 2020, the FDA added GalSafe pigs to the theoretical menu by declaring them safe for both human consumption and medical use. A minor edit to their DNA means that GalSafe pigs don't produce the alpha-gal sugar on the surfaces of their cells. People with alpha-gal syndrome, which is also known as "mammalian meat allergy" and often appears as a consequence of a tick bite, can both eat and receive organs, blood, and other products from GalSafe pigs without fear of anaphylaxis.

Despite these promising signs, the regulatory path for genetically engineered animals remains challenging to navigate. This could change quickly, however, if consumers and regulators (and competitors—a major critic of AquAdvantage salmon has been the Alaskan salmon fishing industry) are willing to be led by the science. Labs around the world are developing genetically modified varieties of domesticated animals to address specific human needs. Fast-growing pigs can increase yields of the most popular meat animal on the planet. Goats that produce anti-diarrheal milk can improve human health in regions where other animals fare poorly. Heat-tolerant cattle can thrive in parts of the world where climate change is driving temperatures upward. Genetic engineering can also reduce the global burden of infectious diseases that cause high mortality among domesticated animals and threaten human health. To that end, labs are developing pigs that are resistant to African swine fever, cattle that can't get mad cow disease, and chickens that can't transmit bird flu either to one another or to people.

Synthetic biology can also create domesticated plants and animals that combat environmental pollution and climate change. Although the Canadian Enviropig project officially ended in 2012, frozen Enviropig sperm could be used to revive the project in a more biotechnology-friendly world, where phosphorus-digesting Enviropigs both save money for pig farmers and reduce eutrophication of the watersheds that serve pig farms. While work with animals remains slowed by regulatory hurdles, significant progress is being made with plants. Scientists from the Salk Institute's Harnessing Plants Initiative, for example, are using synthetic biology to optimize plants' abilities to capture and store carbon by increasing production of suberin, a decay-resistant, carbon-rich protein found in plant roots. The initiative's genetically engineered varieties, which they call IdealPlants™, grow larger and deeper roots than standard plants and therefore sequester more carbon into the soil. They plan to engineer this trait into six of the most common crop plants—corn, canola, soybean, rice, wheat, and cotton—with the goal to harness the world's agricultural system as a weapon to fight climate change.

As synthetic biologists continue to learn how to tweak the genomes of plants and animals to make more, better, and different products, as natural and social scientists refine approaches to evaluate risks of our new biotechnologies, and as practitioners and community members develop methods to deploy these biotechnologies on farms and in forests, we as a global society will become increasingly accustomed to using the tools of synthetic biology to shape our world. Our paradoxical relationship with genetic engineering will be resolved out of necessity. We can't both maintain the comfortable randomness of evolution and at the same time propel our world toward a defined future. If we want enough food to feed 9 or 10 billion people as well as breathable air, drinkable water, and biodiverse habitats, then we need more control over evolution. We need to direct evolution so that species adapt more quickly to today's world. And we must do so in a way that allows everyone, and not only those with the most privilege, access to our biotechnologies. This will require embracing community involvement and cultural differences and moving forward together as a global society. Our survival may depend on it.

THE FUTURE OF US

In October 2018, twin girls Lulu and Nana were born prematurely but without fanfare in a hospital in Shenzhen, China. One month later, their birth was announced by He Jiankui, a biophysicist at Shenzhen's Southern University of Science and Technology, at the Second International Summit on Human Genome Editing in Hong Kong. This time, the announcement of the girls' birth was received with fanfare. It was not, however, the positive reception that He anticipated.

Until November 2018, He had been a relatively minor player in the gene editing scene. He trained first at the University of Science and Technology of China and then at Rice University in Texas, where he earned a PhD in biophysics. He then returned to China and founded a DNA sequencing start-up, Direct Genomics, in Shenzhen.

Several leading scientists and bioethicists in the gene editing field knew He from time that he also spent in San Francisco's Bay area, but none foresaw what He and his research team would eventually do.

During the few years leading up to his announcement, He had shown growing interest in modifying human embryos using gene editing technologies. His research, though, appeared to be restricted to animal embryos or, more controversially, nonviable human embryos, although He was not the first to report such experiments. Some of the experts with whom He interacted in the Bay Area suspected that his long-term plan was to modify human embryos destined for pregnancy. None, however, suspected that He had actually gone through with it, until he announced to them—via email—that the babies had already been born.

While He intended to announce the twins' birth to the rest of the world during the 2018 gene editing summit, he lost control of the story a few days before the meeting. He knew the announcement would cause a stir and was prepared for it. He created YouTube videos to answer what he anticipated would be the most common questions and hired a publicist to help him manage the press. Then, three days before the meeting, Antonio Regalado, a reporter with the MIT *Technology Review*, discovered a new registration of a human gene editing trial on a Chinese website and broke the story. The gene editing community went into a state of disappointed semi-shock. Most believed that it was inevitable that a rogue scientist would eventually create living gene-edited humans, but no one expected that it would be this scientist, this moment, those edits.

Where He had expected fame and praise, he instead received international condemnation. Over the next several months, He lost his university job and was forced out of his leadership role in the company that he cofounded. He eventually went to jail.

The world knows little more about the twins than was revealed in 2018, and nothing at all about a third gene-edited child that was born during the summer of 2019. What we do know comes from leaked copies of He's unpublished manuscript describing his first experiment. The data were, as a warning, a mess.

The experiment began as part of a fertility program. The twins' parents wanted to have a child, but their father was HIV positive, which limited their access to fertility treatment. By enrolling in He's program, they and similarly affected couples would have access to the standard practice of sperm washing, which eliminates the chance that the mother or child will become infected by the virus. He, however, intended to provide the families enrolled in his program with additional protection against HIV. After the embryos were conceived, he would use CRISPR to alter genomes in a way known to protect people from HIV infection. Whether the twins' parents knew about or understood the risks of the additional gene editing procedure is unknown. In fact, the experiment went through such limited ethical review that it's impossible to find out how many people knew what was happening, including whether the hospitals or doctors involved in the IVF procedure knew that they were implanting gene-edited embryos.

After the embryos had grown to somewhere between 200 and 300 cells, He's team removed several of these cells for genome sequencing. These sequencing data reveal several important details of the experiments. First, the edits made to the twins' genomes were not identical to those known to impede HIV-1 infection. The mutation circulating within the human population is a short deletion of DNA that deactivates the CCR5 receptor on human T cells. This deletion blocks the entry of HIV molecules to these cells. People who have two copies of this mutation are less susceptible to HIV infection than are people with one or no copies of the mutation. In one of the twins' genomes, both copies of the CCR5 gene were edited, but the edits were different on the two chromosomes and neither copy mimicked exactly the circulating protective variant. In the other twin's genome, only one copy was edited, and again the edit was different from any other circulating variant. It would there-fore have been impossible to know when the IVF was performed whether the girls would have any protection against HIV infection. Excerpts from the leaked manuscript published in MIT *Technology Review* indicate that He's team learned the specific details of the twins' genome edits before implantation had taken place. Although the embryos could have been frozen in this state and experiments performed to assess the efficacy and safety of these new mutations, He's team forged ahead.

Second, both embryos were mosaics—not all their cells had identical genome sequences. Genomic mosaicism is a known prob-lem when gene editing embryos because the cells have already begun to divide and differentiate into different types of cells and tissues. If the CRISPR editing machinery isn't delivered into each cell, or if different edits are made in different cells, then parts of the body descended from different early cells will differ from one another in their DNA sequences. Additionally, because the few cells for which genomic data were generated were removed before implantation, these particular cells would not become part of the developing child. There is no way to assess how many embryonic cells were edited or whether there were any other, off-target edits to any of the cells that became part of the child's body until after the child was born. The leaked data show that He's team knew about one off-target edit in one of the twins' genomes, but because it did not occur within a part of the genome with a known function, they decided that it would probably have no effect on the child.

Third, He's justification for such an extreme form of medical intervention falls flat. He claimed during his presentation at the summit that he followed community-imposed guidelines intended to regulate human germ line editing. These guidelines were created in 2017 by a panel of scientists and ethicists and recommended that edits only be made to genes that cause disease, that experiments be optimized on animal models prior to engaging in work with human tissues, and that ethical oversight be provided by appropriate scientific and governmental agencies. He's experiment met none of these criteria. HIV is not a genetic disease to be cured by deactivating CCR5; rather the edits were intended to alter an otherwise healthy embryo as a potentially preventative measure. He never tested the specific edits that he made on animal models despite having the opportunity to do so. His ethical oversight appears to have been an afterthought, given that he registered his experiment with the relevant authorities after the twins were born. Rather than intending to solve an otherwise intractable medical problem, He seems most likely to have designed and performed his gene editing experiment so that he could become famous.

He Jiankui's announcement of the birth of gene-edited twins invigorated the gene editing community's already strong objection to human germ line editing. The off-target edits and mosaicism of the twins' embryos confirmed what experts already knew: CRISPR technology is still too poorly optimized to edit human embryos destined for pregnancies, and doing so is reckless and endangers human life. For many, the work also crossed an ethical red line, or at least fell into the ethical gray area between a potentially acceptable treatment and a much more ethically problematic enhancement.

Why do most of us oppose using biotechnology to enhance ourselves? One can imagine a future in which potential parents choose from among the many advantageous mutations that humans around the world harbor in their genomes and combine them using gene editing, creating human embryos that are superbly adapted to performing particular tasks or to living in extreme environments, just as people have done for dogs and cattle. The problem with manipulating human evolution in this way is the inequity that necessarily follows. People don't mind that boxers are stronger than miniature poodles because we intend for boxers and miniature poodles to fill different human niches. Humans, however, are not intended to fill niches but are allowed to decide for themselves how they want to live their lives. We deplore, and fear the misuse of, any technology that might increase the inequities that already exist among us.

But then we get used to them. Forty-five years ago, IVF technology was feared and deplored, but today IVF is celebrated because it makes having a child a possibility for couples unable to conceive on their own. Genetic testing has been adopted across the world to assess the chance that a couple might conceive a child with a genetic disease. Fertility clinics in the United States, Europe, and China now offer DNA sequencing of embryos prior to implantation. These clinics screen embryos for genetic diseases but also allow parents to select traits like biological sex and eye color. Databases containing hundreds of thousands of human genomes have made it possible to link genes to growing numbers of traits, including traits like height, skin pigmentation, and compulsiveness. Fertility clinics can access these data to further characterize each embryo, providing potential parents with detailed assessments of what an embryo's genes predict about their future child's health, looks, intelligence, and behavior. Designing babies by coupling genome sequencing with trait selection is just steps away from designing babies by coupling genome sequencing with DNA manipulation. Large steps, of course, with significant and unsolved technical barriers, but only a few of them.

Another fear may be that once we allow human germ line editing, the results will be too enticing to stop. If we could know with certainty that our child will be smart, attractive, and athletic, why would we not choose to make them so? Once we learn what genes to tweak, what will stop us from tweaking them, other than an outright abolition of the technology?

Opposition to human germ line editing may fade just as opposition to IVF has faded. When and how this might occur probably depend on what happens with the technology over the next several years to decades. As of now, scientists and bioethicists have called for a global moratorium on gene editing human embryos intended for pregnancy, and the International Commission on the Clinical Use of Human Germline Genome Editing is developing revised recommendations to guide human gene editing research, this time with explicit instructions listing precisely what experiments and ethical oversights are required. This does not, however, rule out the possibility that someone somewhere will ignore these recommendations and proceed with editing human embryos intended for pregnancies. Should this happen, there will probably be renewed outcries, renewed calls for increased oversight, renewed worries about a future where designer babies are born with traits that give them an unfair advantage in life. But we will also be slightly less surprised than we were in 2018, and slightly more inured to the idea of gene-edited humans.

We also may be compelled to adopt riskier technologies by crises that emerge from outside our own actions. The COVID-19 pandemic that emerged in China in late 2019 and then spread rapidly throughout the world killed millions of people, disrupted global economies, and overwhelmed health care systems. Because the virus spread between people via aerosol droplets, it also forced us to rethink nearly every part of our society. Schools shuttered. Children were deprived of playdates with their friends. Elderly people, who were among the most susceptible to severe disease, were denied physical interaction with their peers and their families. For nearly an entire year, people everywhere were told to isolate from everyone other than those in their immediate households, taking away the communities and connections that define us as humans. Depression and anxiety increased among people of all age groups. Mental health experts around the world reported increased rates of substance abuse and suicidal ideation. With no vaccine in easy reach, if it had been possible in early 2020 to turn to our new technologies for a solution, would we have hesitated to do so?

The virus responsible for the COVID-19 pandemic, SARS-CoV-2, belongs to a family of viral respiratory diseases called severe acute respiratory syndrome, or SARS. SARS-CoV-2 is a type of virus known as a coronavirus (CoV). Most coronaviruses cause mild and relatively unfrightening illnesses like the common cold. Sometimes, however, coronaviruses are more deadly. The first SARS outbreak happened when a coronavirus jumped into humans from civet cats, which probably got the virus themselves from bats. That outbreak lasted from November 2002 until July 2003, an eight-month window during which 8,098 people in 26 countries were infected and 774 people died. The first SARS outbreak did not become a global pandemic, thanks to quick efforts by international teams of doctors and epidemiologists. It was spread mostly by sick people in health care settings, which made finding and quarantining infected or potentially infected people relatively simple. SARS-CoV-2 was harder to stop because COVID-19 spreads between people who aren't sick enough to know that they are infected.

When SARS-CoV-2 emerged in China sometime during late 2019, we were not prepared, although we probably should have been. It's not as if a pandemic was unexpected. As the number of humans on the planet grows, so does our proximity to one another and to other animals, and so will the chance that new zoonotic diseases (diseases that evolve in an animal host and then jump into humans) emerge. When the pandemic began, however, the scientific community turned its focus toward finding a way to stop SARS-CoV-2. In only a few months, scientists learned enormously more than was known before about how the human immune system responds to infection by SARS and other viruses. Some of what was learned pointed to therapies that improved patient outcomes. Other information was used to develop vaccines that were fast-tracked through the approval process. Scientists also discovered that some people who contracted SARS-CoV-2 were less likely than others to become seriously ill because of genetic variants that they inherited from their parents, similarly to how people who inherit two copies of the CCR5 variant inherit immunity to HIV-1.

COVID-19 is not the deadliest infectious disease currently circulating nor was it the deadliest of pandemics that humans have experienced. But it was the first pandemic that we characterized genetically in real time. We watched the virus evolve, detected the emergence of more virulent strains, and tracked the spread of these strains across the globe. We learned to control COVID-19 using existing interventions: quarantine, therapeutics, and vaccines. But what if that had failed? What if the virus had been more deadly, or spread more quickly? Would we then have turned our attention to the protective genetic variants that we discovered and to the tools at our disposal to edit our own DNA? Would this have been the inflection point—when the need to save ourselves overruled our ethical opposition to altering our own evolutionary path?

I don't doubt that we will eventually turn our engineering powers on ourselves. But when we do, it won't be to chase an unknown and potentially better human. This is not, after all, how evolution works. There is no evolutionary path toward imagining genetic variation that might make one person more fit than another person in some unknown and undescribed future. Instead, we will find ourselves in a moment in which we are forced to decide whether to act or to let nature take its course. Perhaps it will be a time besieged by pandemic disease, in which we've learned that some among us have a genetic variant that makes us less fit in the environment of that moment. Natural selection would remove that variant from our population in a way that we would find unconscionable, in particular when we have the means to avoid that end. And so we will choose a different outcome. We will turn to our technologies to overrule evolution. The once unthinkable ethical breach will become the only imaginable ethical choice. Saving human lives will be our Turkish delight.

IT IS UP TO US

Humans have been manipulating the organisms around them for tens of thousands of years. We began by hunting them, driving some to extinction and others to scarcity, leaving communities and eco-systems to reshuffle in our wake. When we discovered that extinction of our prey was not inevitable, we changed our behavior. We finessed our hunting strategies and began making deliberate choices to sustain prey populations. We learned to transform these animals, as well as the cereals and fruits that we gathered, into better versions of themselves. We brought these animals and plants closer to our settlements and learned to breed the best with the best to create something even better. Our lives improved, and our populations grew. But as we took over increasingly more of the planet, species that we'd taken for granted started to go extinct and the land and water that sustained us deteriorated. So we changed our behavior again. We established rules to protect wild species and wild places. We chose where wild species lived, what they ate, even which among them bred. This left more room for us, and so we intensified our tinkering. We realized that one bull could father thousands of calves, and discovered that the best cattle could be cloned. We engineered crops to be resistant to the pests that we allowed to dominate and to the diseases that we allowed to proliferate. We even figured out how to bring lost lineages back to life. Over the last 50,000 years, we transformed the plants and animals with which we share our planet into lineages that are exquisitely adapted to today's world, where the dominant evolutionary force is us.

We are, however, a different kind of evolutionary force. Evolution is a random walk through experimental space. Evolution does not evaluate risk before making decisions about which breeding experiments to perform, but we do. Evolution doesn't care what the next generation looks like or even if the next generation survives, but we do. Evolution is not guiding horses or wheat or cattle or bison to any particular fate, but we are. And this is where the contradiction in our opposition to our biotechnologies is laid bare. We resist biotechnologies precisely because they give us the control over evolution that humans have continuously worked to achieve. Evolution will not get us to a future that we predetermine. Our biotechnologies, however, can.

The decisions that we make over the next decades will determine our own fate and the fate of other species, perhaps far into the future. We can choose to take advantage of our technologies as they develop, to use synthetic biology to make even more with less, to protect wild species and wild spaces, and to do so in a sustainable way. Or we can reject our new biotechnologies and follow this same path anyway, just more slowly and with less success.

Biotechnologies can be frightening, in particular when they are new. We still have lots of work to do to make our technologies safe, to learn how to assess risk, to collaborate at a global scale. But biotechnologies also give us reason to be hopeful. The world is changing, and people and animals and ecosystems are suffering. Biotechnologies give us the power to help. We can change the evolutionary trajectories of species destined to become extinct. We can clean up our trash and make our farms more efficient. We can cure diseases that afflict us and other species. We can create and sustain a world in which wild species thrive in natural spaces and where people are healthy and happy and decidedly in charge.

"ENDLESS FORMS MOST BEAUTIFUL AND WONDERFUL"

Let's return to that early spring day in Yellowstone National Park. The rustling wind and distant gurgling of a river are the only audible sounds as we watch a herd of bison graze a field of freshly sprouted grass. It is a tranquil and comforting scene. These animals and this space are here because of decisions that people like us made. Sure, humans twice hunted bison nearly to extinction, but a herd survived here in Yellowstone, protected by rules and strategies that we designed to keep them safe. These bison are of course different from their ancestors. They don't face the same challenges of finding food or a mate or avoiding predators that their ancestors did. Some of these bison may even have cattle ancestry that traces back to the aurochs of the Levant. Here in the park, though, they are simply bison, following each other, grunting, snorting, and chewing their way through life.

As I stare, pondering the ancestry and fate of these majestic animals, I'm reminded of the last line of Darwin's *Origin of Species*, which I realize describes perfectly today's human-dominated world. This place and the things that live here are perhaps only wildish. But they are also forms most beautiful, which have been, and continue to be, evolved.

I feel a jab to my lower back as a small foot kicks the driver's seat from the row behind. "Can we go now?" my boys whine in unison, bored with my silent musing and no longer interested in the bison, which, by this time, have disappeared into the horizon. I turn around in my seat to look at them and grin. The area between them in our car is covered in trash—plastic wrappers, empty juice boxes, cracker crumbs, and a smattering of nuts, orange peels, and squashed raisins—a garbage pile that reflects thousands of years of genetic tinkering.

I turn back around, start the car, and adjust my glasses. As we drive toward the park's exit, I call out the names of the wildflowers that I recognize, hoping to distract my kids from their boredom: wild strawberry, phlox, bitterroot, oxeye daisies, glacier lilies, buttercups, balsamroot. Some of these evolved here, others evolved elsewhere and were introduced, perhaps accidentally but perhaps not. The grasses, too, are a mixture of native and exotic species. Maybe they are also interbreeding, intertwining their evolutionary histories and futures just as bison and cattle have, thanks to us. In the backseat, the boys are quiet, lost in their own thoughts and ignoring me, and I smile to myself. This is the way our natural world is, and this is the way our natural world has always been.

And it is most wonderful.

Acknowledgments

I NEED TO THANK LOTS OF PEOPLE FOR HELPING TO MAKE *LIFE AS We Made It* happen. Foremost I need to thank those who helped me to get my facts straight: Ross MacPhee, Grant Zazula, and Paul Koch for their paleontological expertise; Geoff Bailey, Isabelle Winder, Ed Green, and Mindy Zeder for their expertise in early human history, archaeology, and domestication; Alison Van Eenennaam for making sure that my descriptions of genetic engineering and gene editing in agriculture were clear and correct, and also for lighting up my Twitter DMs with photos of cows and newborn calves; Oliver Ryder, Stewart Brand, Kevin Esvelt, Ben Novak, and Ryan Phelan for their knowledge of new biotechnologies for conservation. I would also like to thank several people for feedback that helped to improve earlier drafts: Chris Vollmers, who was willing to read any chapter that loosely involved caves and also came up with the final title; Rachel Meyer, who also provided cocktails when necessary; Matt Schwartz, who needed something to do other than obsessively check the coronavirus statistics; my stepdad Tony Ezzell (of the Palm Coast Ezzells), who read the book to take his mind off the fact that his gym was closed because of the pandemic; and Sarah Crump, Katie Moon, Russ Corbett-Detig, and Shelbi Russell, who tolerated listening to me read paragraphs out loud as I tried to resolve unresolvable clunkiness. Without the contributions of everyone listed here, the book would not be what it is.

I would like to thank Eleanor Jonas, who taught both of my sons when they were in second grade, for allowing me to visit her classroom and talk about mammoths so many years ago. She's probably relieved that I left out the part about the ancient moose poop. And I would also like to thank everyone in the University of California, Santa Cruz, Paleogenomics Lab for dealing like stars with the pandemic slowdowns to their research and for their patience with me as I tried my best to fit everything, including writing a book, into my working-from-home experience.

Speaking of patience, I definitely need to thank my family: my partner, Ed, who read one chapter of this book as an expert reader and swears that he will one day have time to read the whole thing (and also, maybe one day, the mammoth book), and my children, James and Henry, who said that I need to dedicate every book that I write to them because of how much time the process takes away from listening to their enthralling stories about adventures in Minecraft. I am also grateful to those who made it possible for me to focus on writing despite that my house has been bustling with online teaching and meetings and elementary school (and Minecraft) since the world shut down. Without the energy and love that David Capel, Victoria Nobles, and Linda Naranjo invest in James and Henry, there is little chance that I could have finished this book or anything else during the pandemic.

Lastly, I would like to thank my editor, T. J. Kelleher, for ushering me through this process, as well as everyone else at Basic

Books who helped to transform my text into an actual book, and my agent, Max Brockman, for pushing me to believe that I could actually write it.

Now, I'm off to find out whether I can import a genetically modified American chestnut tree into California.

Bibliography and References

THE ANNOTATED LIST OF REFERENCES BELOW IS BY NO MEANS exhaustive. My intention with this list is to direct the reader to resources with which they can explore more deeply the major topics of the preceding chapters. I have tried where possible to point to articles that review the primary scientific literature in an accessible way, or to articles written explicitly for nonscientists. In some cases, in particular where I cite in the text certain data sets or publications, I include the reference in which these data were published.

PROLOGUE: PROVIDENCE
Annotated Bibliography
Among the many potential benefits of primordial germ cell transfer, which I mentioned in the context of chickens that lay eggs that hatch into ducklings (Liu *et al.* 2012), is that common breeds or species can become surrogate parents for rarer breeds or species that are challenging to breed in captivity. Common chicken breeds, for example, have been used to rear offspring of rarer breeds (Woodcock *et al.* 2019). This technology has also been successful in fish (Yoshizaki and Yazawa 2019) and has enormous potential for conservation and aquaculture.

The technical description of the Enviropig can be found in Forsberg *et al.* (2003). Block (2018) describes Enviropigs as well as several other transgenic animals that have thus far not surmounted the hurdles of regulatory approval in a way that may appeal to a less technical audience.

Disagreement about whether biodiversity would be better served by leaving nature alone or intervening more aggressively has deep roots in biodiversity conservation. One side is led by E. O. Wilson, who presents his argument for allowing nature space to recover in *Half-Earth* (2016). On the other side are those of us who believe that it is too late to imagine any portion of the planet not impacted by us and instead we must embrace our role as planetary curators. Two of my favorite books that lay out this argument are Stewart Brand's *Whole Earth Discipline* (2009) and Emma Marris's *Rambunctious Garden* (2011).

The original report of the survey of Aotorean attitudes toward biotechnological solutions for biodiversity conservation can be found in Taylor *et al.* (2017a), and a discussion of the implications of this research is presented in Taylor *et al.* (2017b).

References

Bloch S. 2018. Hornless Holsteins and Enviropigs: The genetically engineered animals we never knew. *The Counter*. https://thecounter.org/transgenesis-gene-editing-fda-aquabounty/.

Brand S. 2009. *Whole Earth Discipline: An Ecopragmatist Manifesto*. New York: Viking Penguin.

Forsberg CW, Phillips JP, Golovan SP, Fan MZ, Meidinger RG, Ajakaiye A, Hilborn D, Hacker RR. 2003. The Enviropig physiology, performance, and contribution to nutrient management advances in a regulated environment: The leading edge of change in the pork industry. *Journal of Animal Science* 81: E68–E77.

Liu C, Khazanehdari KA, Baskar V, Saleen S, Kinne J, Wernery Y, Chang I-K. 2012. Production of chicken progeny (*Gallus gallus domesticus*) from interspecies germline chimeric duck (*Anas domesticus*) by primordial germ cell transfer. *Biology of Reproduction* 86: 1–8.

Marris E. 2011. *Rambunctious Garden: Saving Nature in a Post-Wild World*. New York: Bloomsbury.

Taylor HR, Dussex N, van Heezik Y. 2017a. Bridging the conservation genetics gap by identifying barriers to implementation for conservation practitioners. *Global Ecology and Conservation* 10: 231–242.

Taylor HR, Dussex N, van Heezik Y. 2017b. De-extinction needs consultation. *Nature Ecology and Evolution* 1: 198.

Wilson, EO. 2016. *Half-Earth: Our Planet's Fight for Life*. New York: Liveright.

Woodcock ME, Gheyas AA, Mason AS, Nandi S, Taylor L, Sherman A, Smith J, Burt DW, Hawken R, McGrew MJ. 2019. Reviving rare chicken breeds using genetically engineered sterility in surrogate host birds. *Proceedings of the National Academy of Sciences* 116: 20930–20937.

Yoshizaki G, Yazawa R. 2019. Application of surrogate broodstock technology in aquaculture. *Fisheries Science* 85: 429–437.

CHAPTER 1: BONE MINING
Annotated Bibliography

Reviews of the potential of plant (Zazula *et al.* 2003) and animal (Shapiro and Cooper 2003) remains preserved in the frozen soils of Yukon, Canada, for reconstructing ancient ecosystems underscore the potential of the Klondike goldfields as records of the tumultuous history of the Pleistocene ice ages. Froese *et al.* (2009) explain in relatively nontechnical terms how volcanic ash layers provide a chronology for these events.

Much has been written about the history of bison in North America. Rinella (2009) is among the most thorough historical accounts, even if he does poke fun at the straddling-the-Atlantic accent that I'd adopted by the end of my tenure in Oxford (and have, I think, since lost). Geist (1996) provides fascinating detail about the relationship between Native American people and bison on the midcontinent, and Guthrie (1990) explores the impact of bison on the Beringian ecosystem and the relationships between bison forms in Beringia and elsewhere.

The nineteenth-century decline in bison is well documented by Hornaday (1889), and how this decline impacted North American ecosystems is explored by Yong (2018). Details of the nineteenth- and early twentieth-century efforts to save bison are preserved in various reports of the American Bison Society from that period, which can be found in the society's online archives.

Early twentieth-century attempts to breed bison and cattle were largely unsuccessful (Goodnight 1914) but have nonetheless left traces of cattle ancestry in most bison herds (Halbert and Derr 2009). Derr *et al.* (2012) assess the physical consequences of cattle ancestry for the bison of Santa Catalina Island.

My work reconstructing bison history from ancient DNA began with inferring when their populations grew and declined during the last ice age (Shapiro *et al.* 2004). Later, we recovered DNA from the Chi'jee's Bluff bison foot and a long-horned bison skull from Snowmass, Colorado, to determine when bison first entered North America (Froese *et al.* 2017) and documented their recovery after the peak of the last ice age (Heintzman *et al.* 2016). Our excavations at Snowmass, Colorado (Johnson *et al.* 2014), were featured in an episode of PBS's *NOVA* (Grant 2012).

The first successful amplification of ancient DNA from preserved skin of a quagga is described in Higuchi *et al.* (1984). Subsequent high-profile claims of DNA from dinosaurs and more ancient remains have been debunked as researchers developed and implemented protocols to avoid contamination and validate their results (reviewed in Gilbert *et al.* 2005 and

Shapiro and Hofreiter 2014). Early rules for the field were set out by Cooper and Poinar (2000).

References

Cooper A, Poinar HN. 2000. Ancient DNA: Do it right or not at all. *Science* 289: 1139.

Derr JN, Hedrick PW, Halbert ND, Plough L, Dobson LK, King J, Duncan C, Hunter DL, Cohen ND, Hedgecock D. 2012. Phenotypic effects of cattle mitochondrial DNA in American bison. *Conservation Biology* 26: 1130–1136.

Froese DG, Stiller M, Heintzman PD, Reyes AV, Zazula GD, Soares AER, Meyer M, Hall E, Jensen BKL, Arnold L, MacPhee RDE, Shapiro B. 2017. Fossil and genomic evidence constrains the timing of bison arrival in North America. *Proceedings of the National Academy of Sciences* 114: 3457–3462.

Froese DG, Zazula GD, Westgate JA, Preece SJ, Sanborn PT, Reyes AV, Pearce NJG. 2009. The Klondike goldfields and Pleistocene environments of Beringia. *GSA Today* 19: 4–10.

Geist V. 1996. *Buffalo Nation: History and Legend of the North American Bison*. Stillwater, MN: Voyageur Press.

Gilbert MTP, Bandelt HJ, Hofreiter M, Barnes I. 2005. Assessing ancient DNA studies. *Trends in Ecology and Evolution* 20: 541–544.

Goodnight C. 1914. My experience with bison hybrids. *Journal of Heredity* 5: 197–199.

Grant E. 2012. *Ice Age Death Trap*. NOVA, PBS. www.pbs.org/video/nova-ice -age-death-trap/.

Halbert ND, Derr JN. 2007. A comprehensive evaluation of cattle introgression into US federal bison herds. *Journal of Heredity* 98: 1–12.

Heintzman PD, Froese DG, Ives JW, Soares AER, Zazula GD, Letts B, Andrews TD, Driver JC, Hall E, Hare G, Jass CN, MacKay G, Southon JR, Stiller M, Woywitka R, Suchard MA, Shapiro B. 2016. Bison phylogeography constrains dispersal and viability of the "Ice Free Corridor" in western Canada. *Proceedings of the National Academy of Sciences* 113: 8057–8063.

Higuchi R, Bowman B, Freiberger M, Ryder OA, Wilson AC. 1984. DNA sequences from the quagga, an extinct member of the horse family. *Nature* 312: 282–284.

Hornaday WT. 1889. *Extermination of the American Bison*. In: Smithsonian Institution USNM, editor. Report of the National Museum: Government Printing Office, pp. 369–548.

Johnson KR, Miller IM, Pigati JS. 2014. The Snowmastodon Project. *Quaternary Research* 82: 473–476.

Rinella S. 2009. *American Buffalo: In Search of a Lost Icon*. New York: Spiegel & Grau.

Shapiro B, Cooper A. 2003. Beringia as an Ice Age genetic museum. *Quaternary Research* 60: 94–100.

Shapiro B, Drummond AJ, Rambaut A, Wilson MC, Matheus PE, Sher AV, Pybus OG, Gilbert MT, Barnes I, Binladen J, Willerslev E, Hansen AJ, Baryshnikov GF, Burns JA, Davydov S, Driver JC, Froese DG, Harington CR, Keddie G, . . . Cooper A. 2004. Rise and fall of the Beringian steppe bison. *Science* 306: 1561–1565.

Shapiro B, Hofreiter M. 2014. A paleogenomic perspective on evolution and gene function: New insights from ancient DNA. *Science* 343: 1236573.

Yong E. 2018 November 18. What America lost when it lost the bison. *The Atlantic*.

Zazula GD, Froese DG, Schweger CE, Mathewes RW, Beaudoin AB, Telka AM, Harington CR, Westgate JA. 2003. Ice-age steppe vegetation in East Beringia. *Nature* 423: 603.

CHAPTER 2: ORIGIN STORY
Annotated Bibliography

Ernst Mayr (1942) introduced the biological species concept, although other species concepts have also found favor. Coyne and Orr's book, *Speciation* (2004), provides an in-depth review of species concepts and our evolving understanding of speciation.

Humphrey and Stringer (2018) and Stringer and McKie (2015) offer comprehensive and accessible tours of what fossils reveal about the origin of our lineage, although readers with a shorter attention span may prefer the more concise version in Stringer (2016). Pattison (2020) is another excellent option, in particular for readers also interested in the interpersonal conflicts in paleoanthropology. Lieberman (2014) explores pre-hominin evolution and in particular the evolution of bipedalism. Antón *et al.* (2014) explore the interaction between aridity in Africa and diversification of early *Homo*. Hublin *et al.* (2017) describe the 315,000-year-old modern human fossils from Jebel Irhoud, Morocco.

The first description of *Homo naledi* is in Berger *et al.* (2015). Instructions for 3D printing *H. naledi* elements are on the MorphoSource (https://morphosource.org) Rising Star Project portal.

Although the archaeological record includes some signs of a sudden shift in human behavior (see Mellars 2006 for a summary), most paleoanthropologists believe that human behavioral evolution was a slow and complex process (McBrearty and Brooks 2000). Wurz (2012) provides an overview of the debate that will be accessible to most readers.

Ancient DNA research on Neanderthals began in 1997 (Krings *et al.* 1997), but it was not until Green *et al.* (2010) that these data revealed the deeply intertwined histories of our two lineages. That humans often exchanged genes with our archaic cousins has been reaffirmed many times since 2010, including with ancient DNA from Neanderthals (Prüfer *et al.* 2014, Hajdinjak *et al.* 2018, Mafessoni *et al.* 2020), Denisovans (Reich *et al.* 2010, Meyer *et al.* 2012, Sawyer *et al.* 2015), and hybrids of Neanderthals and Denisovans (Slon *et al.* 2018), as well as from analyses of archaic DNA that persists in ancient and living modern humans (Fu *et al.* 2016, Browning *et al.* 2018). The wider geographic range inhabited by Denisovans was first reported from protein sequences isolated from fossils on the Tibetan Plateau (Chen *et al.* 2019) and later confirmed with ancient DNA isolated directly from sediments on the cave floor (Zhang *et al.* 2020). Meyer *et al.* (2019) report nuclear DNA sequences from fossils recovered from Sima de los Huesos Cave in Spain.

Today, all or at least most of us have small amounts of DNA in our genomes that we inherited from admixture with our archaic cousins (Vernot and Akey 2015, Chen *et al.* 2020). Racimo *et al.* (2015) discuss some of the ways in which DNA passed into the human lineages through admixture with our archaic cousins has impacted our own species' evolutionary history. What all this means to our own evolution is explored in accessible detail by Wei-Hass (2020).

The work growing and analyzing brain organoids with genomes edited to contain the archaic version of NOVA1 is described in Trujillo *et al.* (2021).

References

Antón S, Potts D, Aiello LC. 2014. Early evolution of *Homo*: An integrated biological perspective. *Science* 345: 1236828.

Berger LR, Hawks J, de Ruiter DJ, Churchill SE, Schmid P, Delezene LK, Kivell TL, Garvin HM, Williams SA, DeSilva JM, Skinner MM, Musiba CM, Cameron N, Holliday TW, Harcourt-Smith W, Ackermann RR, Bastir M, Bogin B, Bolter D, . . . Laird MF. 2015. *Homo naledi*, a new species of the genus *Homo* from the Dinaledi Chamber, South Africa. *eLife* 4: e09560.

Browning SR, Browning BL, Zhou Y, Tucci S, Akey JM. 2018. Analyses of human sequence data reveals two pulses of archaic Denisovan admixture. *Cell* 173: 53–61.e9.

Chen F, Welker F, Shen CC, Bailey SE, Bergmann I, Davis S, Xia H, Wang H, Fischer R, Freidline SE, Yu TL, Skinner MM, Stelzer S, Dong G, Fu Q, Dong G, Wang J, Zhang D, Hublin JJ. 2019. A late Middle Pleistocene mandible from the Tibetan Plateau. *Nature* 569: 409–412.

Chen L, Wolf AB, Fu W, Li L, Akey JM. 2020. Identifying and interpreting apparent Neanderthal ancestry in African individuals. *Cell* 180: 677–687.e16.

Coyne JA, Orr HA. 2004. *Speciation*. Sunderland, MA: Sinauer.

Fu Q, Posth C, Hajdinjak M, Petr M, Mallick S, Fernandes D, Furtwängler A, Haak W, Meyer M, Mittnik A, Nickel B, Peltzer A, Rohland N, Slon V, Talamo S, Lazaridis I, Lipson M, Mathieson I, . . . Pääbo S, Reich D. 2016. The genetic history of Ice Age Europe. *Nature* 534: 200–205.

Green RE, Krause J, Briggs AW, Maricic T, Stenzel U, Kircher M, Patterson N, Li H, Zhai W, Fritz MH, Hansen NF, Durand EY, Malaspinas AS, Jensen JD, Marques-Bonet T, Alkan C, Prüfer K, Meyer M, Burbano HA, . . . Pääbo S. 2010. A draft sequence of the Neandertal genome. *Science* 328: 710–722.

Hajdinjak M, Fu Q, Hübner A, Petr M, Mafessoni F, Grote S, Skoglund P, Narasimham V, Rougier H, Crevecoeur I, Semal P, Soressi M, Talamo S, Hublin JJ, Gušić I, Kućan Ž, Rudan P, Golovanova LV, . . . Pääbo S, Kelso J. 2018. Reconstructing the genetic history of late Neanderthals. *Nature* 555: 652–656.

Hublin JJ, Ben-Ncer A, Bailey SE, Freidline SE, Neubauer S, Skinner MM, Bergmann I, Le Cabec A, Benazzi S, Harvati K, Gunz P. 2017. New fossils from Jebel Irhoud, Morocco, and the pan-African origin of *Homo sapiens*. *Nature* 546: 289–292.

Huerta-Sánchez E, Jin X, Asan, Bianba Z, Peter BM, Vinckenbosch N, Liang Y, Yi X, He M, Somel M, Ni P, Wang B, Ou X, Huasang, Luosang J, Cuo ZX, Li K, Gao G, Yin Y, . . . Nielsen R. 2014. Altitude-adaptation in Tibetans caused by introgression of Denisovan-like DNA. *Nature* 512: 194–197.

Humphrey L, Sringer C. 2018. *Our Human Story*. London: Natural History Museum.

Krings M, Stone A, Schmitz RW, Krainitzki H, Stoneking M, Pääbo S. 1997. Neandertal DNA sequences and the origin of modern humans. *Cell* 90: 19–30.

Liberman D. 2014. *The Story of the Human Body: Evolution, Health, and Disease*. New York: Random House.

Mafessoni F, Grote S, de Filippo C, Slon V, Kolobova KA, Viola B, Markin SV, Chintalapati M, Peyrégne S, Skov L, Skoglund P, Krivoshapkin AI, Derevianko AP, Meyer M, Kelso J, Peter B, Prüfer K, Pääbo S. 2020. A high-coverage Neandertal genome from Chagyrskaya Cave. *Proceedings of the National Academy of Sciences* 117: 15132–15136.

Mayr, E. (1942) *Systematics and the Origin of Species*. New York: Columbia University Press.

McBrearty S, Brooks AS. The revolution that wasn't: A new interpretation of the origin of modern human behavior. *Journal of Human Evolution* 39: 453–563.

Mellars P. 2006. Why did modern human populations disperse from Africa ca. 60,000 years ago? A new mode. *Proceedings of the National Academy of Sciences* 103: 9381–9386.

Meyer M, Kircher M, Gansauge MT, Li H, Racimo F, Mallick S, Schraiber JG, Jay F, Prüfer K, de Filippo C, Sudmant PH, Alkan C, Fu Q, Do R, Rohland N, Tandon A, Siebauer M, Green RE, Bryc K, . . . Pääbo S. 2012. A high-coverage genome sequence from an archaic Denisovan individual. *Science* 338: 222–226.

Meyer M, Arsuaga JL, de Filippo C, Nagel S, Aximu-Petri A, Nickel B, Martínez I, Gracia A, Bermúdez de Castro JM, Carbonell E, Viola B, Kelso J, Prüfer K, Pääbo S. 2016. Nuclear DNA sequences from Middle Pleistocene Sima de los Huesos hominins. *Nature* 531: 504–507.

Patterson K. 2020. *Fossil Men: The Quest for the Oldest Skeleton and the Origins of Humankind.* New York: HarperCollins.

Prüfer K, de Filippo C, Grote S, Mafessoni F, Korlević P, Hajdinjak M, Vernot B, Skov L, Hsieh P, Peyrégne S, Reher D, Hopfe C, Nagel S, Maricic T, Fu Q, Theunert C, Rogers R, Skoglund P, Chintalapati M, . . . Pääbo S. 2017. A high-coverage Neandertal genome from Vinfija Cave in Croatia. *Science* 358: 655–658.

Prüfer K, Racimo F, Patterson N, Jay F, Sankararaman S, Sawyer S, Heinze A, Renaud G, Sudmant PH, de Filippo C, Li H, Mallick S, Dannemann M, Fu Q, Kircher M, Kuhlwilm M, Lachmann M, Meyer M, Ongyerth M, . . . Pääbo S. 2014. The complete genome sequence of a Neanderthal from the Altai Mountains. *Nature* 505: 43–49.

Reich D, Green RE, Kircher M, Krause J, Patterson N, Durand EY, Viola B, Briggs AW, Stenzel U, Johnson PL, Maricic T, Good JM, Marques-Bonet T, Alkan C, Fu Q, Mallick S, Li H, Meyer M, Eichler EE, . . . Pääbo S. 2010. Genetic history of an archaic hominin group from Denisova Cave in Siberia. *Nature* 468: 1053–1060.

Sawyer S, Renaud G, Viola B, Hublin JJ, Gansauge MT, Shunkov MV, Derevianko AP, Prüfer K, Kelso J, Pääbo S. 2015. Nuclear and mitochondrial DNA sequence from two Denisovan individuals. *Proceedings of the National Academy of Sciences* 112: 15696–15700.

Slon V, Mafessoni F, Vernot B, de Filippo C, Grote S, Viola B, Hajdinjak M, Peyrégne S, Nagel S, Brown S, Douka K, Higham T, Kozlikin MB, Shunkov MV, Derevianko AP, Kelso J, Meyer M, Prüfer K, Pääbo S. 2018. The genome of the offspring of a Neanderthal mother and a Denisovan father. *Nature* 561: 113–116.

Stringer C, Makie R. 2015. *African Exodus: The Origins of Modern Humanity.* New York: Henry Holt & Co.

Stringer C. 2016. The origin and evolution of *Homo sapiens. Philosophical Transactions of the Royal Society of London, Series B* 371: 20150237.

Trujillo CA, Rice ES, Schaefer NK, Chaim IA, Wheeler EC, Madrigal AA, Buchanan J, Preissl S, Wang A, Negraes PD, Szeto R, Herai RH, Huseynov A, Ferraz MSA, Borges FdS, Kihara AH, Byrne A, Marin M, Vollmers C, . . . Muotri AR. 2021. Reintroduction of archaic variant of *NOVA!* in cortical organoids alters neurodevelopment. *Science* 381: eaax2537.

Vernot B. Akey JM. 2015. Complex history of admixture between modern humans and Neandertals. *American Journal of Human Genetics* 96: 448–453.

Wei-Haas M. 2020 January 30. You may have more Neanderthal DNA than you think. *National Geographic.*

Wurz S. The transition to modern behavior. *Nature Education Knowledge* 3: 15.

Zhang D, Xia H, Chen F, Li B, Slon V, Cheng T, Yang R, Jacobs Z, Dai Q, Massilani D, Shen X, Wang J, Feng X, Cao P, Yang MA, Yao J, Yang J, Madsen DB, Han Y, . . . Fu Q. 2020. Denisovan DNA in Late Pleistocene sediments from Baishiya Karst Cave on the Tibetan Plateau. *Science* 370: 584–587.

CHAPTER 3: BLITZKRIEG
Annotated Bibliography

Evidence for and against humans as the major cause of extinctions over the last 50,000 years has been reviewed by many authors (see Barnosky *et al.* 2004, Koch and Barnosky 2006, Stuart 2014). Ceballos *et al.* (2017) discuss the ongoing extinction crisis and its ties to human population growth and land-use change.

Stuart and Lister (2012) document the decline of the woolly rhinoceros. Counts of woolly rhinoceros remains at archaeological sites in Eurasia are reported in Lorenzen *et al.* (2011). Kosintev *et al.* (2019) review the extinction of Siberian unicorns.

Shapiro *et al.* (2004) first used ancient DNA and a population genetics approach to reconstruct changes in population size over time of an extinct species, in this case bison. A similar approach was subsequently used to infer past dynamics of mammoths (Barnes *et al.* 2007, Palkopoulou *et al.* 2013, Chang *et al.* 2017), muskoxen (Campos *et al.* 2010), woolly rhinoceroses (Lorenzen *et al.* 2011, Lord *et al.* 2020), saber-toothed cats (Paijmans *et al.* 2017), caribou (Lorenzen *et al.* 2001, Kuhn *et al.* 2010), horses (Lorenzen *et al.* 2011), bears (Stiller *et al.* 2010, Edwards *et al.* 2011), lions (Barnett *et al.* 2009), and others.

Our work using sedimentary ancient DNA to determine when and why mammoths became extinct on St. Paul Island, Alaska, is presented in Graham *et al.* (2016). Rogers and Slatkin (2017) show from nuclear genomic data that the last surviving Wrangel Island mammoths were declining due to the effects of inbreeding.

An early wave of human dispersal out of Africa is supported by genetic evidence of hybrid Neanderthals in Germany and Belgium (Peyrégne *et al.*

2019) and fossils attributed to *Homo sapiens* that date to more than 70,000 years ago from China (Liu *et al.* 2015) and Israel (Hershkovitz *et al.* 2018). Dating analyses of the Madjedbebe archaeological site in Australia are presented in Clarkson *et al.* (2017). Sedimentary records of the onset of environmental change in southeastern Australia are from van der Kaars *et al.* (2017). Miller *et al.* (2016) present evidence of human predation on *Genyornis* in Australia.

The timing and route of human dispersal out of Africa and across the planet as interpreted from ancient DNA are summarized by Reich (2016). Heintzman *et al.* (2016) use ancient bison DNA to exclude the ice-free corridor as a potential dispersal route for the earliest humans to colonize the American midcontinent. Matisoo-Smith and Robins (2004) use DNA from rats to reconstruct the timing and ordering of the peopling of the Pacific Islands.

Allentoft *et al.* (2014) show that humans were responsible for the extinction of moa in New Zealand. Langley *et al.* (2020) explore the sustained interaction between humans and their prey in Sri Lanka.

The death of Lonely George on New Year's Day 2019 is reported by Wilcox (2019).

References

Allentoft ME, Heller R, Oskam CL, Lorenzen ED, Hale ML, Gilbert MT, Jacomb C, Holdaway RN, Bunce M. 2014. Extinct New Zealand megafauna were not in decline before human colonization. *Proceedings of the National Academy of Sciences* 111: 4922–4927.

Barnes I, Shapiro B, Kuznetsova T, Sher A, Guthrie D, Lister A, Thomas MG. 2007. Genetic structure and extinction of the woolly mammoth. *Current Biology* 17: 1072–1075.

Barnett R, Shapiro B, Ho SYW, Barnes I, Burger J, Yamaguchi N, Higham T, Wheeler HT, Rosendhal W, Sher AV, Baryshnikov G, Cooper A. 2009. Phylogeography of lions (*Panthera leo*) reveals three distinct taxa and a Late Pleistocene reduction in genetic diversity. *Molecular Ecology* 18: 1668–1677.

Barnosky AD, Koch PL, Feranec RS, Wing SL, Shabel AB. 2004. Assessing the causes of Late Pleistocene extinctions on the continents. *Science* 306: 70–75.

Campos P, Willerslev E, Sher A, Axelsson E, Tikhonov A, Aaris-Sørensen K, Greenwood A, Kahlke R-D, Kosintsev P, Krakhmalnaya T, Kuznetsova T, Lemey P, MacPhee RD, Norris CA, Shepherd K, Suchard MA, Zazula GD, Shapiro B, Gilbert MTP. 2010. Ancient DNA analysis excludes humans as the driving force behind Late Pleistocene musk ox (*Ovibos moschatus*) population dynamics. *Proceedings of the National Academy of Sciences* 107: 5675–5680.

Ceballos G, Ehrlich PR, Dirzo R. 2017. Biological annihilation via the on-going sixth mass extinction signaled by vertebrate population losses and declines. *Proceedings of the National Academy of Sciences* 114: E6089–E6096.

Chang D, Knapp M, Enk J, Lippold S, Kircher M, Lister A, MacPhee RDE, Widga C, Czechowski P, Sommer R, Hodges E, Stümpel N, Barnes I, Dalén L, Derevianko A, Germonpré M, Hillebrand-Voiculescu A, Constantin S, Kuznetsova T, . . . Shapiro B. 2017. The evolutionary and phylogeographic history of woolly mammoths: A comprehensive mitogenomic analysis. *Scientific Reports* 7: 44585.

Clarkson C, Jacobs Z, Marwick B, Fullagar R, Wallis L, Smith M, Roberts RG, Hayes E, Lowe K, Carah X, Florin SA, McNeil J, Cox D, Arnold LJ, Hua Q, Huntley J, Brand HEA, Manne T, Fairbairn A, . . . Pardoe C. 2017. Human occupation of northern Australia by 65,000 years ago. *Nature* 547: 306–310.

Edwards CJE, Suchard MA, Lemey P, Welch JJ, Barnes I, Fulton TL, Barnett R, O'Connell TC, Coxon P, Monaghan N, Valdiosera C, Lorenzen ED, Willerslev E, Baryshnikov GF, Rambaut A, Thomas MG, Bradley DG, Shapiro B. 2011. Ancient hybridization and a recent Irish origin for the modern polar bear matriline. *Current Biology* 21: 1–8.

Graham RW, Belmecheri S, Choy K, Cullerton B, Davies LH, Froese D, Heintzman PD, Hritz C, Kapp JD, Newsom L, Rawcliffe R, Saulnier-Talbot E, Shapiro B, Wang Y, Williams JW, Wooller MJ. 2016. Timing and cause of mid-Holocene mammoth extinction on St. Pal Island, Alaska. *Proceedings of the National Academy of Sciences* 113: 9310–9314.

Heintzman PD, Froese DG, Ives JW, Soares AER, Zazula GD, Letts B, Andrews TD, Driver JC, Hall E, Hare G, Jass CN, MacKay G, Southon JR, Stiller M, Woywitka R, Suchard MA, Shapiro B. 2016. Bison phylogeography constrains dispersal and viability of the "Ice Free Corridor" in western Canada. *Proceedings of the National Academy of Sciences* 113: 8057–8063.

Hershkovitz I, Weber GW, Quam R, Duval M, Grün R, Kinsley L, Ayalon A, Bar-Matthews M, Valladas H, Mercier N, Arsuaga JL, Martinón-Torres M, Bermúdez de Castro JM, Fornai C, Martín-Francés L, Sarig R, May H, Krenn VA, Slon V, . . . Weinstein-Evron M. 2018. The earliest modern humans outside of Africa. *Science* 359: 456–459.

Koch PL, Barnosky AD. 2006. Late Quaternary extinctions: State of the debate. *Annual Reviews of Ecology and Evolution* 37: 215–250.

Kosintsev P, Mitchell KJ, Devièse T, van der Plicht J, Kuitems M, Petrova E, Tikhonov A, Higham T, Comeskey D, Turney C, Cooper A, van Kolfschoten T, Stuart AJ, Lister AM. 2019. Evolution and extinction of the giant rhinoceros *Elasmotherium sibricum* sheds light on Late Quaternary megafaunal extinctions. *Nature Ecology and Evolution* 3: 31–38.

Kuhn TS, McFarlane K, Groves P, Moers AO, Shapiro B. 2010. Modern and ancient DNA reveal recent partial replacement of caribou in the southwest Yukon. *Molecular Ecology* 19: 1312–1318.

Langley MC, Amano N, Wedage O, Deraniyagala S, Pathmalal MM, Perera N, Boivin N, Petraglia MD, Roberts P. 2020. Bows and arrows and complex symbolic displays 48,000 years ago in the South Asian Tropics. *Science Advances* 6: eaba3831.

Liu W, Martinón-Torres M, Cai YJ, Xing S, Tong HW, Pei SW, Sier MJ, Wu XH, Edwards RL, Cheng H, Li YY, Yang XX, de Castro JM, Wu XJ. 2015. The earliest unequivocally modern humans in southern China. *Nature* 526: 696–699.

Lord E, Dussex N, Kierczak M, Díez-Del-Molino D, Ryder OA, Stanton DWG, Gilbert MTP, Sánchez-Barreiro F, Zhang G, Sinding MS, Lorenzen ED, Willerslev E, Protopopov A, Shidlovskiy F, Fedorov S, Bocherens H, Nathan SKSS, Goossens B, van der Plicht J, . . . Dalén L. 2020. Pre-extinction demographic stability and genomic signatures of adaptation in the woolly rhinoceros. *Current Biology* 5: 3871–3879.

Lorenzen ED, Nogués-Bravo D, Orlando L, Weinstock J, Binladen J, Marske KA, Ugan A, Borregaard MK, Gilbert MT, Nielsen R, Ho SY, Goebel T, Graf KE, Byers D, Stenderup JT, Rasmussen M, Campos PF, Leonard JA, Koepfli KP, . . . Willerslev E. 2011. Species-specific responses of Late Quaternary megafauna to climate and humans. *Nature* 479: 359–364.

Matisoo-Smith E, Robins JH. 2004. Origins and dispersals of Pacific peoples: Evidence from mtDNA phylogenies of the Pacific rat. *Proceedings of the National Academy of Sciences* 101: 9167–9172.

Miller G, Magee J, Smith M, Spooner N, Baynes A, Lehman S, Fogel M, Johnston H, Williams D, Clark P, Florian C, Holst R, DeVogel S. 2016. Human predation contributed to the extinction of the Australian megafaunal bird *Genyornis newtoni* ~47ka. *Nature Communications* 7: 10496.

Paijmans JLA, Barnett R, Gilbert MTP, Zepeda-Mendoza ML, Reumer JWF, de Vos J, Zazula G, Nagel D, Baryshnikov GF, Leonard JA, Rohland N, Westbury MV, Barlow A, Hofreiter M. 2017. Evolutionary history of saber-toothed cats based on ancient mitogenomics. *Current Biology* 27: 3330–3336.e5.

Palkopoulou E, Dalén L, Lister AM, Vartanyan S, Sablin M, Sher A, Edmark VN, Brandström MD, Germonpré M, Barnes I, Thomas J. 2013. Holarctic genetic structure and range dynamics in the woolly mammoth. *Proceedings of the Royal Society of London, Series B* 280: 20131910.

Peyrégne S, Slon V, Mafessoni F, de Filippo C, Hajdinjak M, Nagel S, Nickel B, Essel E, Le Cabec A, Wehrberger K, Conard NJ, Kind CJ, Posth C, Krause J, Abrams G, Bonjean D, Di Modica K, Toussaint M, Kelso J, . . . Prüfer K. 2019. Nuclear DNA from two early Neandertals reveals 80,000 years of genetic continuity in Europe. *Science Advances* 5: eaaw5873.

Reich D. 2018. *Who We Are and How We Got Here*. Oxford: Oxford University Press.

Rogers RL, Slatkin M. 2017. Excess of genomic defects in a woolly mammoth on Wrangel Island. *PLoS Genetics* 13: e1006601.

Shapiro B, Drummond AJ, Rambaut A, Wilson MC, Matheus PE, Sher AV, Pybus OG, Gilbert MT, Barnes I, Binladen J, Willerslev E, Hansen AJ, Baryshnikov GF, Burns JA, Davydov S, Driver JC, Froese DG, Harington CR, Keddie G, . . . Cooper A. 2004. Rise and fall of the Beringian steppe bison. *Science* 306: 1561–1565.

Stiller M, Baryshnikov G, Bocherens H, Grandal d'Anglade A, Hilpert B, Münzel SC, Pinhasi R, Rabeder G, Rosendahl W, Trinkaus E, Hofreiter M, Knapp M. 2010. Withering away—25,000 years of genetic decline preceded cave bear extinction. *Molecular Biology and Evolution* 27: 975–978.

Stuart AJ. 2014. Late Quaternary megafaunal extinctions on the continents: A short review. *Geological Journal* 50: 338–363.

Stuart AJ, Lister AM. 2012. Extinction chronology of the woolly rhinoceros *Coelodonta antiquitis* in the context of Late Quaternary megafaunal extinctions in northern Eurasia. *Quaternary International* 51: 1–17.

van der Kaars S, Miller GH, Turney CS, Cook EJ, Nürnberg D, Schönfeld J, Kershaw AP, Lehman SJ. 2017. Humans rather than climate the primary cause of Pleistocene megafaunal extinction in Australia. *Nature Communications* 8: 14142.

Wilcox C. 2019 January 8. Lonely George the tree snail dies, and a species goes extinct. *National Geographic*.

CHAPTER 4: LACTASE PERSISTENCE
Annotated Bibliography

Ségurel and Bon (2017) review current understanding of the evolution and functional consequences of lactase persistence mutations in humans. Genetic data have been used to infer the timing of appearance of the lactase persistence mutation and trace its spread across Eurasia (Segruel *et al.* 2020) and Africa (Tishkoff *et al.* 2007). Little evidence has been found of this mutation in early dairying populations (Burger *et al.* 2020).

Zeder (2011) reviews what the archaeological record has revealed so far about the process of animal domestication in the Levant and the dispersal of domesticated animals into Europe. Marshall *et al.* (2014) evaluate the role of directed breeding during early domestication. The three pathways to domestication are summarized in Zeder (2012 and 2015). The syndrome of traits common to domesticated species, coined "domestication syndrome" in the early twentieth century, is described by Wilkins *et al.* (2014).

Schultz *et al.* (2005) compare agriculture in ants to agriculture in humans.

Ancient DNA has contributed to our understanding of where and when many species were domesticated (reviewed in Frantz *et al.* 2020), including dogs (Bergström *et al.* 2020), chickens (Wang *et al.* 2020), and horses (Orlando 2020). The timing and consequences of the expansion of the first horse-herding people are discussed by Haak *et al.* (2015) and de Barros Damgaard *et al.* (2018).

Genomic insights into cattle domestication are presented by Park *et al.* (2015) and Verdugo *et al.* (2019). Pitt *et al.* (2019) debate whether the evidence supports two or three cattle domestication events globally. Archaeological evaluation of aurochs fossils from Dja'ade (Helmer *et al.* 2005) and Çayönü (Hongo *et al.* 2009) are discussed in the broader context of cattle domestication in the Fertile Crescent by Arbuckle *et al.* (2016). Qiu *et al.* (2012) identify regions of the genomes of Tibetan cattle that originated from admixture with local yaks and have enabled these aurochs to survive at high altitude.

Archaeological evidence of dairying includes fats preserved in ceramic vessels in Europe (Evershed *et al.* 2008) and Africa (Grillo *et al.* 2020), as well as protein sequences recovered from archaeological dental plaque (Warinner *et al.* 2014; Charlton *et al.* 2019 review this approach and the discoveries it has enabled).

Hare and Tomasello (2005) explore evidence of social cognition in dogs (see also Hare and Woods 2013). Saito *et al.* (2019) and Vitale and Udell (2019) show that cats, too, evolved traits that suggest social reliance on their human companions.

Recent domestication of marine species is documented by Duarte *et al.* (2007), and Stokstad (2020) evaluates the growing role of biotechnology in aquaculture. Rosner (2014) explores present efforts to domesticate new species of wild plants, including the American potato bean.

Moore and Hasler (2017) review how twentieth-century biotechnologies have advanced cattle husbandry and dairy science. Hansen (2020) discusses some of the ongoing challenges with these technologies, considering why embryo transfer in particular has thus far not reached its promise. Wiggans *et al.* (2017) evaluate the contribution of genomic data to breeding selection in modern cattle.

Bannasch *et al.* (2008) link a mutation in the gene SLC2A9 to the overproduction of uric acid in Dalmatians. Lewis and Mellersh (2019) report declining frequencies of disease in dogs since the adoption of DNA testing.

References

Arbuckle BS, Price MD, Hongo H, Öksüz B. 2016. Documenting the initial appearance of domestic cattle in the eastern Fertile Crescent (northern Iraq and western Iran). *Journal of Archaeological Science* 72: 1–9.

Bannasch D, Safra N, Young A, Kami N, Schaible RS, Ling GV. 2008. Mutations in the *SLC2A9* gene cause hyperuricosuria and hyperuremia in the dog. *PLoS Genetics* 4: e1000246.

Bergström A, Frantz L, Schmidt R, Ersmark E, Lebrasseur O, Girdland-Flink L, Lin AT, Storå J, Sjögren KG, Anthony D, Antipina E, Amiri S, Bar-Oz G, Bazaliiskii VI, Bulatović J, Brown D, Carmagnini A, Davy T, Fedorov S, . . . Skoglund P. 2020. Origins and genetic legacy of prehistoric dogs. *Science* 370: 557–564.

Burger J, Link V, Blöcher J, Schulz A, Sell C, Pochon Z, Diekmann Y, Žegarac A, Hofmanová Z, Winkelbach L, Reyna-Blanco CS, Bieker V, Orschiedt J, Brinker U, Scheu A, Leuenberger C, Bertino TS, Bollongino R, Lidke G, . . . Wegmann D. 2020. Low prevalence of lactase persistence in Bronze Age Europe indicates ongoing strong selection over the last 3,000 years. *Current Biology* 30: 4307–4315.

Charlton S, Ramsøe A, Collins M, Craig OE, Fischer R, Alexander M, Speller CF. 2019. New insights into Neolithic milk consumption through proteomic analysis of dental calculus. *Archaeological and Anthropological Sciences* 11: 6183–6196.

Craig OE, Chapman J, Heron C, Willis LH, Bartosiewicz L, Taylor G, Whittle A, Collins M. Did the first farmers of central and eastern Europe produce dairy foods? *Antiquity* 79: 882–894.

de Barros Damgaard P, Martiniano R, Kamm J, Moreno-Mayar JV, Kroonen G, Peyrot M, Barjamovic G, Rasmussen S, Zacho C, Baimukhanov N, Zaibert V, Merz V, Biddanda A, Merz I, Loman V, Evdokimov V, Usmanova E, Hemphill B, Seguin-Orlando A, . . . Willerslev E. 2018. The first horse herders and the impact of early Bronze Age steppe expansions into Asia. *Science* 360: eaar7711.

Duarte CM, Marbá N, Jolmer M. 2007. Rapid domestication of marine species. *Science* 316: 382–383.

Evershed RP, Payne S, Sherratt AG, Copley MS, Coolidge J, Urem-Kotsu D, Kotsakis K, Ozdoğan M, Ozdoğan AE, Nieuwenhuyse O, Akkermans PM, Bailey D, Andeescu RR, Campbell S, Farid S, Hodder I, Yalman N, Ozbaşaran M, . . . Burton MM. 2008. Earliest date for milk use in the Near East and southeastern Europe linked to cattle herding. *Nature* 455: 528–531.

Felius M. 2007. *Cattle Breeds: An Encyclopedia.* Pomfret, VT: Trafalgar Square Publishing.

Frantz LAF, Bradley DG, Larson G, Orlando L. 2020. Animal domestication in the era of ancient genomics. *Nature Reviews Genetics* 21: 449–460.

Grillo KM, Dunne J, Marshall F, Prendergast ME, Casanova E, Gidna AO, Janzen A, Karega-Munene, Keute J, Mabulla AZP, Robertshaw P, Gillard T, Walton-Doyle C, Whelton HL, Ryan K, Evershed RP. Molecular and isotopic evidence for milk, meat, and plants in prehistoric eastern African herder food systems. *Proceedings of the National Academy of Sciences* 117: 9793–9799.

Hansen PJ. 2020. The incompletely fulfilled promise of embryo transfer in cattle—why aren't pregnancy rates greater and what can we do about it? *Journal of Animal Science* 98: skaa288.

Hare B, Tomasello M. 2005. Human-like social skills in dogs? *Trends in Cognitive Science* 9: 439–444.

Hare B, Woods V. 2013. *The Genius of Dogs: How Dogs Are Smarter Than You Think*. New York: Dutton.

Helmer D, Gourichon L, Monchot H, Peters J, Saña Seguí M. 2005. Identifying early domestic cattle from pre-pottery Neolithic sites on the Middle Euphrates using sexual dimorphism. In: Vigne J-D, Peters J, Helmer D, editors. *New Methods and the First Steps of Mammal Domestication*. Oxford: Oxbow Books, pp. 86–95.

Hongo H, Pearson J, Öksüz B, Ígezdi G. 2009. The process of ungulate domestication at Çayönü, southeastern Turkey: A multidisciplinary approach focusing on *Bos* sp. and *Cervus elaphus*. *Anthropozoologica* 44: 63–78.

Kistler L, Montenegro A, Smith BD, Gifford JA, Green RE, Newsom LA, Shapiro B. 2014. Trans-oceanic drift and the domestication of African bottle gourds in the Americas. *Proceedings of the National Academy of Sciences* 111: 2937–2941.

Lewis TW, Mellersh CS. 2019. Changes in mutation frequency of eight Mendelian inherited disorders in eight pedigree dog populations following introduction of a commercial DNA test. *PLoS One* 14: e0209864.

Librado P, Fages A, Gaunitz C, Leonardi M, Wagner S, Khan N, Hanghøj K, Alquraishi SA, Alfarhan AH, Al-Rasheid KA, Der Sarkissian C, Schubert M, Orlando L. 2016. The evolutionary origin and genetic makeup of domestic horses. *Genetics* 204: 423–434.

Marshall FB, Dobney K, Denham T, Capriles JM. 2014. Evaluating the roles of directed breeding and gene flow in animal domestication. *Proceedings of the National Academy of Sciences* 111: 6153–6158.

Moore SG, Hasler JF. 2017. A 100-year review: Reproductive technologies in dairy science. *Journal of Dairy Science* 100: 10314–10331.

Orlando L. 2020. The evolutionary and historical foundation of the modern horse: Lessons from ancient genomics. *Annual Reviews of Genetics* 54: 561–581.

Park SDE, Magee DA, McGettigan PA, Teasdale MD, Edwards CJ, Lohan AJ, Murphy A, Braud M, Donoghue MT, Liu Y, Chamberlain AT, Rue-Albrecht K, Schroeder S, Spillane C, Tai S, Bradley DG, Sonstegard TS, Loftus B, MacHugh DE. Genome sequencing of the extinct Eurasian wild aurochs, *Boss primigenius*, illuminate the phylogeography and evolution of cattle. *Genome Biology* 16: 234.

Pitt D, Sevane N, Nicolazzi EL, MacHugh DE, Park SDE, Colli L, Martinez R, Bruford MW, Orozco-terWengel P. 2019. Domestication of cattle: Two or three events? *Evolutionary Applications* 2019: 123–136.

Qiu Q, Zhang G, Ma T, Qian W, Wang J, Ye Z, Cao C, Hu Q, Kim J, Larkin DM, Auvil L, Capitanu B, Ma J, Lewin HA, Qian X, Lang Y, Zhou R, Wang L, Wang K, . . . Liu J. 2012. The yak genome and adaptation to life at high altitude. *Nature Genetics* 44: 946–949.

Rosner H. 2014 June 24. How we can tame overlooked wild plants to feed the world. *Wired*.

Saito A, Shinozuka K, Ito Y, Hasegawa T. 2019. Domestic cats (*Felis catus*) discriminate their names from other words. *Scientific Reports* 9: 5394.

Schaible RH. 1981. The genetic correction of health problems. *The AKC Gazette*.

Schultz T, Mueller U, Currie C, Rehner S. 2005. Reciprocal illumination: A comparison of agriculture in humans and in fungus-growing ants. In: Vega F, Blackwell M, editors. *Ecological and Evolutionary Advances in Insect-Fungal Associations*. Oxford: Oxford University Press, pp. 149–190.

Ségurel L, Bon C. 2017. On the evolution of lactase persistence in humans. *Annual Review of Genomics and Human Genetics* 18: 297–319.

Ségurel L, Guarino-Vignon P, Marchi N, Lafosse S, Laurent R, Bon C, Fabre A, Hegay T, Heyer E. 2020. Why and when was lactase persistence selected for? Insights from Central Asian herders and ancient DNA. *PLoS Biology* 18: e30000742.

Stokstad E. 2020. Tomorrow's catch. *Science* 370: 902–905.

Tishkoff SA, Reed FA, Ranciaro A, Voight BF, Babbitt CC, Silverman JS, Powell K, Mortensen HM, Hirbo JB, Osman M, Ibrahim M, Omar SA, Lema G, Nyambo TB, Ghori J, Bumpstead S, Pritchard JK, Wray GA, Deloukas P. 2007. Convergent adaptation of human lactase persistence in Africa and Europe. *Nature Genetics* 39: 31–40.

Verdugo MP, Mullin VE, Scheu A, Mattiangeli V, Daly KG, Maisano Delser P, Hare AJ, Burger J, Collins MJ, Kehati R, Hesse P, Fulton D, Sauer EW, Mohaseb FA, Davoudi H, Khazaeli R, Lhuillier J, Rapin C, Ebrahimi S, . . . Bradley DG. 2019. Ancient cattle genomics, origins, and rapid turnover in the Fertile Crescent. *Science* 365: 173–176.

Vitale KR, Udell MAR. 2019. The quality of being sociable: The influence of human attentional state, population, and human familiarity on domestic cat sociability. *Behavioral Processes* 145: 11–17.

Wang MS, Thakur M, Peng MS, Jiang Y, Frantz LAF, Li M, Zhang JJ, Wang S, Peters J, Otecko NO, Suwannapoom C, Guo X, Zheng ZQ, Esmailizadeh A, Hirimuthugoda NY, Ashari H, Suladari S, Zein MSA, Kusza S, . . . Zhang YP. 2020. 863 genomes reveal the origin and domestication of chicken. *Cell Research* 30: 693–701.

Warinner C, Hendy J, Speller C, Cappellini E, Fischer R, Trachsel C, Arneborg J, Lynnerup N, Craig OE, Swallow DM, Fotakis A, Christensen RJ, Olsen JV, Liebert A, Montalva N, Fiddyment S, Charlton S, Mackie M, Canci A, . . . Collins MJ. 2014. Direct evidence of milk consumption from ancient human dental calculus. *Scientific Reports* 4: 7104.

Wiggans GR, Cole JB, Hubbard SM, Sonstegard TS. 2017. Genomic selection in dairy cattle: The USDA experience. *Annual Reviews of Animal Biosciences* 5: 309–327.

Wilkins AS, Wrangham RW, Fitch WT. 2014. The "domestication syndrome" in mammals: A unified explanation based on neural crest cell behavior and genetics. *Genetics* 197: 795–808.

Zeder M. 2011. The origins of agriculture in the Near East. *Current Anthropology* 54: S221–S235.

Zeder M. 2012. The domestication of animals. *Journal of Archaeological Research* 68: 161–190.

Zeder M. 2015. Core questions in domestication research. *Proceedings of the National Academy of Sciences* 112: 3191–3198.

CHAPTER 5: LAKE COW BACON
Annotated Bibliography

I was first introduced to the story of H.R. 23261 by Jon Mooallem, whose 2013 article about the team of misfits who tried to push the legislation through the US Congress is not to be missed. Greenburg (2014) provides an equally engrossing story of the plight of the passenger pigeon in the decades leading up to its extinction.

We eventually recovered mitochondrial DNA from the leg of the Oxford dodo (Shapiro *et al.* 2002, Soares *et al.* 2016), but our complete dodo nuclear genome is from a different animal that is part of the collection at the Natural History Museum of Denmark. This work is ongoing. Our analyses of the evolutionary history of the passenger pigeon are presented in Murray *et al.* (2018).

Estes *et al.* (2016) explore the relationship among sea otters, kelp forests, and the now extinct Steller's sea cow.

Sagarin and Turnipseed (2012) review how the legal framework of the Public Trust Doctrine is interpreted in conservation today. Wikipedia provides a thorough traipse through the history of the environmental and conservation movements in the United States and elsewhere. A six-part documentary produced by PBS (Duncan *et al.* 2009) describes the political and social background leading to the creation of the US National Park System and is an excellent taste of the prevailing and shifting attitudes toward conservation of the time. The official website of the US Fish and Wildlife

Service includes details of the history of the Endangered Species Act and its various amendments.

Pimm *et al.* (2014) explore the impact of international biodiversity-related conventions on the protection of endangered species. The efficacy of approaches in use today to slow biodiversity loss is reviewed by Johnson *et al.* (2017). Bongaarts (2019) summarizes the key findings of the 2019 report of the United Nations' Intergovernmental Science-Policy Platform on Biodiversity and Ecosystem Services. The full report, which was published on May 9, 2019, can be accessed at the website https://ipbes.net /global-assessment.

The story of the near extinction and genetic rescue of the Florida panther is documented in chapters 4 and 5 of O'Brien (2003). Our genomic analysis of inbreeding in the Florida panther populations is described in Saremi *et al.* (2019).

An edited book by Ruiz and Carlton (2003) reviews both how species become invasive and strategies to control invasive species. Kistler *et al.* (2014) find that bottle gourds dispersed to the Americas by riding currents across the Atlantic Ocean. Russell and Broome (2016) review the ecological effects of rodent removal from islands. Milius (2020) reported the 2016 discovery of murder hornets in a package entering the United States through the San Francisco airport.

References

Bongaarts J. 2019. Summary for policymakers of the global assessment report on biodiversity and ecosystem services of the Intergovernmental Science-Policy Platform on Biodiversity and Ecosystem Services. *Population and Development Review* 45: 680–681.

Duncan D, Burns K, Coyote P, Stetson L, Arkin A, Bosco P, Conway K, Hanks T, Lucas J, McCormick C, Bodett T, Clark T, Guyer M, Jones G, Madigan A, Wallach E, Muir J. 2009. *The National Parks: America's Best Idea*. Arlington, VA: PBS Home Video.

Estes JA, Burdin A, Doak DF. 2016. Sea otters, kelp forests, and the extinction of Steller's sea cow. *Proceedings of the National Academy of Sciences* 113: 880–885.

Greenburg J. 2014. *A Feathered River Across the Sky: The Passenger Pigeon's Flight to Extinction*. New York: Bloomsbury.

Johnson CN, Balmford A, Brook BW, Buettel JC, Galetti M, Guangchun L, Wilmshurst JM. 2017. Biodiversity losses and conservation responses in the Anthropocene. *Science* 356: 270–275.

Kistler L, Montenegro A, Smith BD, Gifford JA, Green RE, Newsom LA, Shapiro B. 2014. Trans-oceanic drift and the domestication of African bottle gourds in the Americas. *Proceedings of the National Academy of Sciences* 111: 2937–2941.

Milius S. 2020 May 29. More "murder hornets" are turning up. Here's what you need to know. *Science News*.

Mooallem J. 2013 December 12. American hippopotamus. *The Atavist*.

Murray GGR, Soares AER, Novak BJ, Schaefer NK, Cahill JA, Baker AJ, Demboski JR, Doll A, Da Fonseca RR, Fulton TL, Gilbert MTP, Heintzman PD, Letts B, McIntosh G, O'Connell BL, Peck M, Pipes M-L, Rice ES, Santos KM, . . . Shapiro B. 2017. Natural selection shaped the rise and fall of passenger pigeon genomic diversity. *Science* 358: 951–954.

O'Brien SJ. 2003. *Tears of the Cheetah and Other Tales from the Genetic Frontier*. New York: Thomas Dunne.

Pimm SL, Jenkins CN, Abell R, Brooks TM, Gittleman JL, Joppa LN, Raven PH, Roberts CM, Sexton JO. 2014. The biodiversity of species and their rates of extinction, distribution, and protection. *Science* 344: 1246752.

Roosevelt T. 2017. "A Book Lover's Holidays in the Open, 1916." In: *Theodore Roosevelt for Nature Lovers: Adventures with America's Great Outdoorsman*. Dawidziak M, editor. Guilford, CT: Lyons Press.

Ruiz GM, Carlton JT. 2003. *Invasive Species: Vectors and Management Strategies*. Washington DC: Island Press.

Russell JC, Broome KG. 2016. Fifty years of rodent eradications in New Zealand: Another decade of advances. *New Zealand Journal of Ecology* 40: 197–204.

Sagarin RD, Turnipseed M. 2012. The Public Trust Doctrine: Where ecology meets natural resources management. *Annual Review of Environment and Resources* 37: 473–496.

Saremi N, Supple MA, Byrne A, Cahill JA, Lehman Coutinho L, Dalén L, Figueiró HV, Johnson WE, Milne HJ, O'Brien SJ, O'Connell BO, Onorato DP, Riley SPD, Sikich JA, Stahler DR, Villetta PMS, Vollmers C, Wayne RK, Eizirik E, . . . Shapiro B. 2019. Puma genomes from North and South America provide insights into the genomic consequences of inbreeding. *Nature Communications* 10: 4769.

Shapiro B, Sibthorpe D, Rambaut A, Austin J, Wragg GM, Bininda-Emonds OR, Lee PL, Cooper A. 2002. Flight of the dodo. *Science* 295: 1683.

Soares AER, Novak B, Haile J, Fjeldså J, Gilbert MTP, Poinar H, Church G, Shapiro B. 2016. Complete mitochondrial genomes of living and extinct pigeons revise the timing of the columbiform radiation. *BMC Evolutionary Biology* 16: 1–9.

Topics of the Times. 1910 April 12. *New York Times*.

Transcript of the presentation of H.R. 23261. 1910. Hearings before the Committee on Agriculture during the second session of the Sixty-first Congress 3. Washington, DC: US Government Printing Office.

Wilson ES. 1934. Personal recollections of the passenger pigeon. *The Auk* 51: 157.

CHAPTER 6: POLLED
Annotated Bibliography

Van Eenennaam *et al.* (2021) review bioengineering technologies and their use in agriculture as well as the opportunities lost because of our reticence and regulatory hurdles. Mendlesohn *et al.* (2003) present an analysis by the US Environmental Protection Agency of risks associated with *Bt* crops. The US National Academies produced a detailed report of the prospects of genetic engineering in agriculture and of scientific, societal, and political challenges that remain (National Academies of Science, Engineering, and Mathematics 2016).

The *Coordinated Framework for Regulation of Biotechnology* in the United States was published in 1986 (Office of Science and Technology Policy 1986) and updated in early January 2017 (Office of the President 2017). The US Food and Drug Administration's 2017 regulatory guidance states that organisms with intentional alterations must be regulated as "new animal drugs" (Food and Drug Administration 2017).

The European Union's definition of a GMO is from Plan and Van den Eede (2010). Laaninen (2019) discusses the rationale for and implications of the 2018 decision by the European Court of Justice that gene-edited organisms fall under the remit of existing European legislation governing GMOs. The global economic implications of this decision are explored by Purnhagen and Wesseler (2021).

Cohen *et al.* (1972) describe the first use of restriction enzymes to join together DNA strands from two different organisms. The quote from an audience member of the 1973 Gordon Conference on Nucleic Acids was reported in Hanna (1991), which also includes details of the experiments leading up to this announcement and the events that followed, including the 1975 meeting in Asilomar, California, and includes commentary by many of the scientists involved with the discussions at the time. Berg *et al.* (1975) summarizes the conclusions of the Asilomar conferences.

Tzifra and Citovsky (2006) review the technologies used to edit the genomes of plants. Kim and Kim (2014) describe the development and application of programmable nucleases for genetic engineering. Doudna and Charpentier (2014) and Knott and Doudna (2018) focus specifically on advances made possible using CRISPR-Cas systems. Reynolds (2019) provides a summary intended for a broader audience of how CRISPR works and what it might be used to accomplish.

The use of antisense technology to control PG activity in tomatoes was first described by Sheehy *et al.* (1988), and the implications of this discovery were reported by Roberts (1988). Details of experiments performed and decisions made by Calgene during its quest for regulatory approval of the Flavr Savr tomato are from Martineau (2001). Calgene published its safety assessment of the Flavr Savr tomato in 1992 (Radenbaugh *et al.* 1992). Searbrook (1993) interviews Calgene officials about the launch of the Flavr Savr tomato. Miller (1993) explores reactions by the media and activist organizations.

Carlson *et al.* (2016) report the birth of hornless dairy cattle using genetically engineered cell lines. The phenotypic and genetic analyses of one of these bull's six calves are presented in Young *et al.* (2020). Norris *et al.* (2020) identify bacterial DNA sequences in the original gene-edited male's genome. The full story of Princess and her siblings (with photos!) is told by Molteni (2019).

The original 2012 study by Séralini and colleagues that presented data purportedly showing a propensity of rats to get cancer when fed a diet of GMO corn was retracted from *Food and Chemical Toxicology* and later republished as Séralini *et al.* (2014). The response of the media to the original study is criticized by Butler (2012). Steinberg *et al.* (2019) could not replicate Séralini's results using larger sample sizes.

Saletan (2015) explores the science behind and controversy surrounding several genetically modified organisms, including genetically modified papaya and Golden Rice. Losey *et al.* (1999) found that *Bt* corn pollen is harmful to monarch butterflies, but their claim was refuted by Sears *et al.* (2001) with data from six large field studies. Tang *et al.* (2012) reported that children fed a serving of Golden Rice received a substantial proportion of the daily recommended dose of vitamin A. Following ensuing controversy (Hvistendahl and Enserink 2012), the article describing the study was retracted. Afedraru (2018) describes the vitamin-fortified, genetically modified banana being developed for release in Uganda. Plans to modify plants in South Africa to survive prolonged periods of drought are discussed by Lind (2017).

Lewis (1992) coined the term "Frankenfoods."

References

Afedraru L. 2018 October 30. Ugandan scientists poised to release vitamin-fortified GMO banana. *Alliance for Science.*

Berg P, Baltimore D, Brenner S, Roblin RO, Singer MF. 1975. Summary statement of the Asilomar conference on recombinant DNA molecules. *Proceedings of the National Academy of Sciences* 72: 1981–1984.

Butler D. 2012. Rat study sparks GM furore. *Nature* 489: 474.

Carlson DF, Lancto CA, Zang B, Kim ES, Walton M, Oldeschulte D, Seabury C, Sonstegard TS, Fahrenkrug SC. 2016. Production of hornless dairy cattle from genome-edited cell lines. *Nature Biotechnology* 34: 479–481.

Cohen SN, Chang ACY, Boyer HW, Helling RB. 1972. Construction of biologically functional bacterial plasmids in vitro. *Proceedings of the National Academy of Sciences* 71: 3240–3244.

Doudna JA, Charpentier E. 2014. The new frontier of genome engineering with CRISPR-Cas9. *Science* 346: 1258096.

Food and Drug Administration. 2017. Guidance for Industry 187 on regulation of intentionally altered genomic DNA in animals. *Federal Register* 82: 12.

Hanna KE, ed. 1991. *Biomedical Politics*. Washington, DC: National Academies Press.

Hvistendahl M, Enserink M. 2012. GM research: Charges fly, confusion reigns over Golden Rice study in Chinese children. *Science* 337: 1281.

Kim H, Kim J-S. 2014. A guide to genome engineering with programmable nucleases. *Nature Reviews Genetics* 15: 321–334.

Knott GJ, Doudna JA. 2018. CRISPR-Cas guides the future of genetic engineering. *Science* 361: 866–869.

Laaninen T. 2019. *New plant-breeding techniques: Applicability of EU GMO rules*. Brussels: European Parliamentary Research Service.

Lewis P. 1992 June 16. Opinion: Mutant foods create risks we can't yet guess. *New York Times*.

Lind P. 2017 March 22. "Resurrection plants": Future drought-resistant crops could spring back to life thanks to gene switch. *Reuters*.

Losey JE, Rayor LS, Carter ME. 1999. Transgenic pollen harms monarch larvae. *Nature* 399: 214.

Martineau B. 2001. *First Fruit: The Creation of the Flavr Savr Tomato and the Birth of Biotech Foods*. New York: McGraw Hill.

Mendelsohn M, Kough J, Vaituzis Z, Matthews K. 2003. Are Bt crops safe? *Nature Biotechnology* 21: 1003–1009.

Miller SK. 1994 May 28. Genetic first upsets food lobby. *New Scientist*.

Molteni M. 2019 October 8. A cow, a controversy, and a dashed dream of more human farms. *Wired*.

National Academies of Sciences, Engineering, and Medicine. 2016. *Genetically Engineered Crops: Experiences and Prospects*. Washington, DC: National Academies Press.

Norris AL, Lee SS, Greenless KJ, Tadesse DA, Miller MF, Lombardi HA. 2020. Template plasmid integration in germline genome-edited cattle. *Nature Biotechnology* 38: 163–164.

Office of Science and Technology Policy. 1986. Coordinated Framework for Regulation of Biotechnology. *Federal Register* 51: 23302.

Office of the President. 2017. Modernizing the Regulatory System for Biotechnology Products: Final Version of the 2017 Update to the Coordinated Framework for the Regulation of Biotechnology. US EPA. www.epa.gov

/regulation-biotechnology-under-tsca-and-fifra/update-coordinated-frame
work-regulation-biotechnology.

Plan D, Van den Eede G. 2010. *The EU Legislation on GMOs: An Overview.*
Brussels: Publications Office of the European Union.

Purnhagen K, Wesseler J. 2021. EU regulation of new plant breeding technologies and their possible economic implications for the EU and beyond.
Applied Economic Perspectives and Policy. https//:doi:10.1002/aepp.13084.

Redenbaugh K, Hiatt W, Martineau B, Kramer M, Sheehy R, Sanders R,
Houck C, Emlay D. 1992. *Safety Assessment of Genetically-Engineered
Fruits and Vegetables: A Case Study of the FLAVR SAVR™ Tomatoes.* Boca
Raton: CRC Press.

Reynolds M. 2019 January 20. What is CRISPR? The revolutionary gene-
editing tech explained. *Wired.*

Roberts L. 1988. Genetic engineers build a better tomato. *Science* 241: 1290.

Saletan W. 2015 July 15. Unhealthy fixation. *Slate.*

Searbrook J. 1993 July 19. Tremors in the hothouse. *New Yorker:* 32–41.

Sears MK, Hellmich RL, Stanley-Horn DE, Oberjauser KS, Pleasants JM,
Mattila HR, Siegfried BD, Dively GP. 2001. Impact of *Bt* corn pollen
on monarch butterfly populations: A risk assessment. *Proceedings of the
National Academy of Sciences* 98: 11937–11942.

Séralini GE, Clair E, Mesnage R, Gress S, Defarge N, Malatesta M, Henne-
quin D, de Vendômois JS. 2014. Long term toxicity of a Roundup herbicide and a Roundup-tolerant genetically modified maize. *Environmental
Sciences Europe* 26: 14.

Sheehy R, Kramer M, Hiatt W. 1988. Reduction of polygalacturonase activity
in tomato fruit by antisense RNA. *Proceedings of the National Academy of
Sciences* 85: 8805–8809.

Simon F. 2015 November 26. Jeremy Rifkin: "Number two cause of global
warming emissions? Animal husbandry." *Euractiv.*

Steinberg P, van der Voet H, Goedhart PW, Kleter G, Kok EJ, Pla M, Nadal
A, Zeljenková D, Aláčová R, Babincová J, Rollerová E, Jaďuďová S, Ke-
bis A, Szabova E, Tulinská J, Líšková A, Takácsová M, Mikušová ML,
Krivošíková Z, . . . Wilhelm R. 2019. Lack of adverse effects in subchronic
and chronic toxicity/carcinogenicity studies on the glyphosate-resistant
genetically modified maize NK603 in Wistar Han RCC rats. *Archives of
Toxicology* 93: 1095–1139.

Tang G, Hu Y, Yin S, Wang Y, Dallal GE, Grusak MA, Russell RM. 2012.
β-Carotene in Golden Rice is as good as β-carotene in oil at providing vitamin A to children. *American Journal of Clinical Nutrition* 96:
658–664.

Tzfira T, Citovsky V. 2006. *Agrobacterium*-mediated genetic transformation of
plants: Biology and biotechnology. *Current Opinion in Biotechnology* 17:
147–154.

Van Eenennaam AL, De Figuieredo Silva F, Trott JF, Zilberman D. 2021. Genetic engineering of livestock: The opportunity cost of regulatory delay. *Annual Review of Animal Biosciences* 9: 453–478.

Young AE, Mansour TA, McNabb BR, Owen JR, Trott JF, Brown CT, Van Eenennaam AL. 2020. Genomic and phenotype analyses of six offspring of a genome-edited hornless bull. *Nature Biotechnology* 38: 225–232.

CHAPTER 7: INTENDED CONSEQUENCES
Annotated Bibliography

The title of this chapter, *Intended Consequences*, was inspired by a workshop convened in June 2020 by Revive & Restore to mark the launch of the Intended Consequences Initiative, which aims to encourage innovation in conservation and promote interventions designed to achieve intentional change. Resources from this workshop, including links to relevant readings and a proposed code of practice, are available at https://reviverestore.org/what-we-do/intended-consequences/.

In Shapiro (2015), I provide a step-by-step guide to de-extinction, exploring how existing and future technologies could be used to reintroduce extinct traits to living species or, perhaps, extinct species to living ecosystems. The July/August supplement of the Hastings Center Report presents a series of essays about the ethics, practice, and future of de-extinction as a tool in biodiversity conservation, with an introduction to the special issue by Kaebnick and Jennings (2017).

Yamagata *et al.* (2019) report efforts by Akira Iritani's team to revive 28,000-year-old mammoth nuclei as a first step toward cloning mammoths. Wu-Suk Hwang's team's protocols for cloning deceased dogs are presented in Jeong *et al.* (2020), and Cyranoski (2006) describes Hwang's 2006 trial for fraud, embezzlement, and bioethics violations. Sarchet (2017) discusses George Church's effort to create a mammoth starting with cells grown in dishes in his lab.

Wilmut *et al.* (2015) review the technology, limitations, and prospects of somatic cell nuclear transfer. Borges and Pereira (2019) discuss the potential of cloning for conservation, including when a different species is used as maternal host. Wani *et al.* (2017) report the successful cloning of a Bactrian camel using a domestic camel as a surrogate mother. Madrigal (2013) describes the project to bring the extinct bucardo back to life. The experiments to engineer the cloned bucardo are presented in Folch *et al.* (2009).

Zimov *et al.* (2012) argue that resurrected mammoths will transform the tundra into a productive steppe grassland. Malhi *et al.* (2016) review our current understanding of how megafauna impact ecosystems, with special consideration to the idea of rewilding ecosystems in which megafauna have become extinct.

The Wyoming Game and Fish Department produced an excellent series of videos about the history of the black-footed ferret conservation project that are available on its YouTube channel (www.youtube.com /user/wygameandfish/) and at http://blackfootedferret.org. Dobson and Lyles (2000) describe the initial comeback of the black-footed ferret during the end of the twentieth century. The international collaboration to genetically rescue the black-footed ferret is described on the Revive & Restore website.

Popkin (2020) explores the history of American chestnut trees, their near extinction, and efforts to save them from that fate. The science behind transgenic American chestnut trees is detailed by Powell *et al.* (2019). Newhouse and Powell (2021) present their argument for American chestnut restoration using genetic engineering.

Ferguson (2018) describes the global scale of the problem of vector-borne diseases and reviews approaches to control mosquito populations, including *Wolbachia* and gene drives. Oxitec describes its engineered mosquito, OX513A, in Harris *et al.* (2012); other insects engineered by Oxitec are described on the company website, which also provides links to relevant scientific literature. Evans *et al.* (2019) reported that some OX513A DNA was transferred into local mosquito populations in Brazil. The decision to release OX5034 mosquitoes in the Florida Keys was reported by Bote (2020).

Burt *et al.* (2018) describe the problem of endemic malaria in sub-Saharan Africa and the malaria control strategies that Target Malaria hopes to use to address this problem. Data pertaining to the initial release of sterile male mosquitoes in Bana, Burkina Faso, are on the Target Malaria website.

Scudellari (2019) provides a tour through gene drive technologies that should be accessible to a broad audience. Kevin Esvelt's Mice Against Ticks project is described in Buchtal *et al.* (2019). Noble *et al.* (2019) present the idea of daisy-chain gene drives. Kevin Esvelt also explains different gene drive systems in a series of videos available on the MIT Media Lab's YouTube channel (www.youtube.com/user/mitmedialab). The first successful gene drive in a mammal was reported by Grunwald *et al.* (2019).

References

Borges AA, Pereira AF. 2019. Potential role of intraspecific and interspecific cloning in the conservation of wild animals. *Zygote* 27: 111–117.

Bote J. 2020 August 20. More than 750 million genetically modified mosquitoes to be released into Florida Keys. *USA Today*.

Buchthal J, Weiss Evans S, Lunshof J, Telford SR III, Esvelt KM. 2019. Mice Against Ticks: An experimental community-guided effort to prevent tick-borne disease by altering the shared environment. *Philosophical Transactions of the Royal Society of London Series B* 374: 20180105.

Burt A, Coulibaly M, Crisanti A, Diabate A, Kayondo JK. 2018. Gene drive to reduce malaria transmission in sub-Saharan Africa. *Journal of Responsible Innovation* 5: S66–S80.

Cyranoski D. 2006. Hwang takes the stand at fraud trial. *Nature* 444: 12.

Dobson A, Lyles A. 2000. Black-footed ferret recovery. *Science* 288: 985–988.

Evans BR, Kotsakiozi P, Costa-da-Silva AL, Ioshino RS, Garziera L, Pedrosa MC, Malavasi A, Virginio JF, Capurro ML, Powell JR. 2019. Transgenic *Aedees aegypri* mosquitoes transfer genes into a natural population. *Scientific Reports* 9: 13047.

Ferguson NM. 2018. Challenges and opportunities in controlling mosquito-borne infections. *Nature* 559: 490–497.

Folch J, Cocero MJ, Chesné P, Alabart JL, Domínguez V, Cognié Y, Roche A, Fernández-Arias A, Martí JI, Sánchez P, Echegoyen E, Beckers JF, Bonastre AS, Vignon X. 2009. First birth of an animal from an extinct subspecies (*Capra pyrenaica pyrenaica*) by cloning. *Theriogenology* 71: 1026–1034.

Grunwald HA, Gantz VM, Poplawski G, Xu X-RS, Bier E, Cooper KL. 2019. Super-Mendelian inheritance mediated by CRISPR-Cas9 in the female mouse germline. *Nature* 566: 105–109.

Harris AF, McKemey AR, Nimmo D, Curtis Z, Black I, Morgan SA, Oviedo MN, Lacroix R, Naish N, Morrison NI, Collado A, Stevenson J, Scaife S, Dafa'alla T, Fu G, Phillips C, Miles A, Raduan N, Kelly N, . . . Alphey L. 2012. Successful suppression of a field mosquito population by sustained release of engineered male mosquitoes. *Nature Biotechnology* 30: 828–830.

Jeong Y, Olson OP, Lian C, Lee ES, Jeong YW, Hwang WS. 2020. Dog cloning from post-mortem tissue frozen without cryoprotectant. *Cryobiology* 97: 226–230.

Kaebnick GE, Jennings B. 2017. De-extinction and conservation: An introduction to the special issue "Recreating the wild: De-extinction, technology, and the ethics of conservation." *Hastings Center Report* 47: S2–S4.

Madrigal A. 2013 March 18. The 10 minutes when scientists brought a species back from extinction. *The Atlantic*.

Malhi Y, Doughty CE, Galetti M, Smith FA, Svenning J-C, Terborgh JW. 2016. Megafauna and ecosystem function from the Pleistocene to the Anthropocene. *Proceedings of the National Academy of Sciences* 113: 838–846.

Newhouse AE, Powell WA. 2021. Intentional introgression of a blight tolerance transgene to rescue the remnant population of American chestnut. *Conservation Science and Practice*. https://doi.org/10.1111/csp2.348.

Noble C, Min J, Olejarz J, Buchthal J, Chavez A, Smidler AL, DeBenedictis EA, Church GM, Nowak MA, Esvelt KM. 2019. Daisy-chain gene drives for the alteration of local populations. *Proceedings of the National Academy of Sciences* 116: 8275–8282.

Popkin G. 2020 April 30. Can genetic engineering bring back the American chestnut? *New York Times Magazine*.

Powell WA, Newhouse AE, Coffey V. 2019. Developing blight-tolerant American chestnut trees. *Cold Spring Harbor Perspectives in Biology* 11: a034587.

Sarchet P. 2017 February 16. Can we grow woolly mammoths in the lab? George Church hopes so. *New Scientist*.

Scudellari M. 2019. Self-destructing mosquitoes and sterilized rodents: The promise of gene drives. *Nature* 57: 160–162.

Shapiro B. 2015. *How to Clone a Mammoth: The Science of De-Extinction.* Princeton, NJ: Princeton University Press.

Wani NA, Vettical BS, Hong SB. 2017. First cloned Bactrian camel (*Camelus bactrianus*) calf produced by interspecies somatic cell nuclear transfer: A step towards preserving the critically endangered wild Bactrian camels. *PLoS One* 12: e0177800.

Wilmut I, Bai Y, Taylor J. 2015. Somatic cell nuclear transfer: Origins, the present position and future opportunities. *Philosophical Transactions of the Royal Society Series B* 310: 20140366.

Yamagata K, Nagai K, Miyamoto H, Anzai M, Kato H, Miyamoto K, Kurosaka S, Azuma R, Kolodeznikov II, Protopopov AV, Plotnikov VV, Kobayashi H, Kawahara-Miki R, Kono T, Uchida M, Shibata Y, Handa T, Kimura H, Hosoi Y, . . . Iritani A. 2019. Signs of biological activities of 28,000-year-old mammoth nuclei in mouse oocytes visualized by live-cell imaging. *Scientific Reports* 9: 4050.

Zimov SA, Zimov NS, Tikhonov AN, Chapin FS III. 2012. Mammoth steppe: A high-productivity phenomenon. *Quaternary Science Reviews* 57: 26–45.

CHAPTER 8: TURKISH DELIGHT
Annotated Bibliography

Waltz (2019) explores consumers' willingness to consume genetically engineered foods, including the Impossible Burger. Guy Raz interviews Pat Brown about his transition from academia to purveyor of plant-based foods in a 2020 episode of NPR's *How I Built This*.

Our collaborative project with Gingko Bioworks to reconstitute extinct scents is described by Kiedaisch (2019). Maloney *et al.* (2018) compares the efficacy of recombinant factor C in detecting the presence of toxic amoebae in medicines to that of the protein in horseshoe crab blood that recombinant factor C is intended to replace.

Gong *et al.* (2003) describe the production via genetic engineering of three varieties of glowing zebrafish. Hill *et al.* (2014) found that GloFish presented no additional risks to the environment than non–genetically engineered danio. Broom (2004) reports that scientists from the Roslin Institute in Edinburgh, Scotland, created GFP-expressing pigs and chickens for research. Other transgenic animals discussed in this chapter include the ruby puppy Ruppy (Callaway 2009), the micropig (Standeart 2017), and the double-muscled beagle (Regalado 2015).

Parker (2018) describes Captain Charles Moore's discovery of the Great Pacific Garbage Patch and explores what has been learned about oceanic garbage patches since that discovery. Biello (2008) and Tullo (2019) discuss the future of biodegradable plastics. The potential of microbes and worms to degrade plastic waste is explored by Drahl (2018).

Kwon *et al.* (2020) report genetically engineered tomatoes designed for urban gardens. Conrow (2016) interviews Ron Stotish, CEO of AquaBounty, about AquAdvantage salmon. GalSafe pigs are described by Phelps *et al.* (2003). The Harnessing Plants Initiative's IdealPlants™ technology is described on the Salk Institute website (www.salk.edu/harnessing-plants-initiative/).

He Jiankui's journey to create and then reveal the world's first CRISPR-modified humans is described by Cohen (2019). Regalado (2018) broke the story of the successfully gene-edited babies prior to He's public revelation. Unpublished details of He's experiments were revealed by Regalado (2019). Cyranoski (2020) confirmed the existence of a third CRISPR baby and discusses the impact of He's prison sentence on Chinese research. The community-developed guidelines for the governance of human genome editing research were published by the National Academies of Science, Engineering, and Medicine (2017).

Ellinghause *et al.* (2020) identify genetic variants that confer different risks of severe COVID-19 infection in humans.

References

Biello D. 2008 September 16. Turning bacteria into plastic factories. *Scientific American*.

Broom S. 2004 April 28. Green-tinged farm points the way. *BBC News*.

Callaway E. 2009 April 23. Fluorescent puppy is world's first transgenic dog. *New Scientist*.

Cohen J. 2018 August 1. The untold story of the "circle of trust" behind the world's first gene-edited babies. *Science*.

Conrow J. 2016 June 20. AquaBounty: GMO pioneer. *Alliance for Science*.

Cyranoski D. 2020. What CRISPR-baby prison sentences mean for research. *Nature* 577: 154–155.

Darwin C. 1859. *On the Origin of Species by Means of Natural Selection, or Preservation of Favoured Races in the Struggle for Life.* London: John Murray.

Drahl C. 2018 June 15. Plastics recycling with microbes and worms is further away than people think. *Chemical and Engineering News* 96.

Ellinghaus D et al. (Severe Covid-19 GWAS Group). 2020. Genomewide association study of severe COVID-19 with respiratory failure. *New England Journal of Medicine* 383: 1522–1534.

Gong Z, Wan H, Leng Tay T, Wang H, Chen M, Yan T. 2003. Development of transgenic fish for ornamental and bioreactor by strong expression of fluorescent proteins in the skeletal muscle. *Biochemical and Biophysical Research Communications* 308: 58–63.

Hill JE, Lawson LL, Hardin S. 2014. Assessment of the risks of transgenic fluorescent ornamental fishes to the United States using the Fish Invasiveness Screening Kit (FISK). *Transactions of the American Fisheries Society* 143: 817–829.

Kiedaisch J. 2019 April 16. You can now smell a flower that went extinct a century ago. *Popular Mechanics*.

Kwon C-T, Heo J, Lemmon ZH, Capua Y, Hutton SF, Van Eck J, Park SJ, Lippman ZB. 2020. Rapid customization of Solanaceae fruit crops for urban agriculture. *Nature Biotechnology* 38: 182–188.

Maloney T, Phelan R, Simmons M. 2018. Saving the horseshoe crab: A synthetic alternative to horseshoe crab blood for endotoxin detection. *PLoS Biology* 16: e2006607.

National Academies of Sciences, Engineering, and Medicine. 2017. *Human Genome Editing: Science, Ethics, and Governance.* Washington, DC: National Academies Press.

Parker L. 2018 March 22. The Great Pacific Garbage Patch isn't what you think it is. *National Geographic*.

Phelps CJ, Koike C, Vaught TD, Boone J, Wells KD, Chen SH, Ball S, Specht SM, Polejaeva IA, Monahan JA, Jobst PM, Sharma SB, Lamborn AE, Garst AS, Moore M, Demetris AJ, Rudert WA, Bottino R, Bertera S, . . . Ayares DL. 2003. Production of alpha 1,3-galactosyltransferase-deficient pigs. *Science* 299: 411–414.

Raz G. 2020 May 11. Impossible Foods: Pat Brown. *How I Built This with Guy Raz.* National Public Radio.

Regalado A. 2015 October 19. First gene-edited dogs reported in China. *MIT Technology Review*.

Regalado A. 2018 November 25. Exclusive: Chinese scientists are creating CRISPR babies. *MIT Technology Review*.

Regalado A. 2019 December 3. China's CRISPR babies: Read exclusive excerpts from unseen original research. *MIT Technology Review.*

Standaert M. 2017 July 3. China genomics giant drops plans for gene-edited pets. *MIT Technology Review.*

Tullo AH. 2019 September 8. PHA: A biopolymer whose time has finally come. *Chemical and Engineering News* 97.

Waltz E. 2019. Appetite grows for biotech foods with health benefits. *Nature Biotechnology* 37: 573–575.

Index

BETH SHAPIRO is a professor of evolutionary biology at the University of California, Santa Cruz, and an Investigator at the Howard Hughes Medical Institute. She is the author of *How to Clone a Mammoth*, which won the AAAS/Subaru Prize for Excellence in Science Books. She lives in Santa Cruz, California.